Massimo Stiavelli
From First Light to Reionization

For additional information reagarding this topic, please refer also to the following publications

Phillipps, S.

The Structure and Evolution of Galaxies

2005
ISBN 978-0-470-85507-2

Stahler, S. W., Palla, F.

The Formation of Stars

2004
ISBN 978-3-527-40559-6

Roos, M.

Introduction to Cosmology

2003
ISBN 978-0-470-84910-1

Liddle, A.

An Introduction to Modern Cosmology

2003
ISBN 978-0-470-84835-7

Shore, S. N.

The Tapestry of Modern Astrophysics

2003
ISBN 978-0-471-16816-4

Coles, P., Lucchin, F.

Cosmology
The Origin and Evolution of Cosmic Structure

2002
ISBN 978-0-471-48909-2

Massimo Stiavelli

From First Light to Reionization

The End of the Dark Ages

WILEY-VCH Verlag GmbH & Co. KGaA

The Author

Massimo Stiavelli
Space Telescope Science Inst.
3700, San Martin Dr. Baltimore
MD 21218
USA

All books published by Wiley-VCH are carefully produced. Nevertheless, authors, editors, and publisher do not warrant the information contained in these books, including this book, to be free of errors. Readers are advised to keep in mind that statements, data, illustrations, procedural details or other items may inadvertently be inaccurate.

Library of Congress Card No.: applied for
British Library Cataloguing-in-Publication Data: A catalogue record for this book is available from the British Library.
Bibliographic information published by the Deutsche Nationalbibliothek
The Deutsche Nationalbibliothek lists this publication in the Deutsche Nationalbibliografie; detailed bibliographic data are available on the Internet at http://dnb.d-nb.de.

© 2009 WILEY-VCH Verlag GmbH & Co. KGaA, Weinheim

All rights reserved (including those of translation into other languages). No part of this book may be reproduced in any form by photoprinting, microfilm, or any other means nor transmitted or translated into a machine language without written permission from the publishers. Registered names, trademarks, etc. used in this book, even when not specifically marked as such, are not to be considered unprotected by law.

Typesetting le-tex publishing services oHG, Leipzig
Printing betz-druck GmbH, Darmstadt
Binding Litges & Dopf GmbH, Heppenheim
Cover Spiesz Design, Neu-Ulm

Printed in the Federal Republic of Germany
Printed on acid-free paper

ISBN: 978-3-527-40705-7

Preface

Deciding whether a topic is ready to become textbook material can be a judgement call and, sometimes, a gamble. The latter is especially true when a field is evolving rapidly thanks to experimental and theoretical advances. The study of the highest-redshift frontier has been one of the most exciting areas of astronomy in the last decade. Improved observational tools such as large ground-based telescopes of the 8–10 m class and the Hubble Space Telescope allowed us to push this frontier to redshift 6 and beyond. We are now finally at the point of being able to address the formation of the first stars and to study the objects responsible and the chronology of the reionization of hydrogen. This represents, in a sense, the boundary of classical astronomy. The Universe before reionization and before the formation of the first stars was indeed a very different place. We expect that new facilities such as the Wide-Field Camera 3 to be installed on the Hubble Space Telescope during Servicing Mission 4 and, especially, the James Webb Space Telescope will enable major observational progress in this area and the flood of new observational results will most likely stimulate further theoretical progress.

Given this positive outlook and great expectations of progress one might wonder why it is a good idea to write a textbook on such a fast-evolving topic. The reasons are several. One is that a field that is lively attracts the attention of many observers and theorists and it is likely that many students will work in this field in the coming decade. However, the bulk of the work already done is very significant and – before we started this enterprise – no graduate-level textbook gave an overview of the state-of-the-art on both theory and observations so as to enable young researchers to get quickly up to speed in this field of research. Another reason is that despite the many unknowns, some of the present thinking regarding the formation of the first stars and reionization is deeply rooted in physical arguments so that, even if the details change, many of the general principles should remain applicable. This topic relies on results from many branches of physics ranging from cosmology to atomic physics and chemical reactions, so that providing a simple but reasonably complete overview of these various aspects can be useful not only to students but also to active researchers specialized in other areas of astronomy who may be experts in some of the subareas but not in others.

As an Editor of the Wiley Cosmology series my first thought was to find suitable authors for this book but unfortunately it is hard to find top researchers actively

From First Light to Reionization. Massimo Stiavelli
Copyright © 2009 WILEY-VCH Verlag GmbH & Co. KGaA, Weinheim
ISBN: 978-3-527-40705-7

engaged in a field willing to pause and write a book of this nature. After several unsuccessful attempts I decided to try myself and I do not regret it. Writing this book certainly took a lot of effort but was also interesting and very rewarding and I learned a lot in the process.

I have attempted to capture what appear to be the most solid results and concepts. In the effort to collect the material and the ideas for this book I have profited from discussions with many colleagues including Tom Abel, Mike Fall, Xhiahoui Fan, Harry Ferguson, Zoltan Haiman, Peter Jakobsen, Simon Lilly, Avi Loeb, Colin Norman, Nino Panagia, Paul Shapiro, Mike Shull, Peter Stockman, Rogier Windhorst, Rosie Wyse. In writing the theoretical chapters of this book I have enjoyed daily interactions and discussions with Michele Trenti. A fraction of this work was done during a sabbatical year spent at Johns Hopkins University. My wife, Francesca Boffi, contributed her patience and understanding, which were crucial for the completion of this project.

Baltimore, September 2008 *Massimo Stiavelli*

Contents

Preface *V*

1 Introduction *1*
1.1 First Light and Reionization *1*
1.2 The Cosmological Framework *2*
1.3 Organization of this Book *2*
1.4 Key Observations in this Field *4*

Part 1 Theory *5*

2 The First Stars *7*
2.1 Overview *7*
2.1.1 First Light *7*
2.1.2 Forming the First Stars *8*
2.1.3 The Legacy of the First Stars *9*
2.2 Before the First Stars *9*
2.2.1 Recombination and Residual Ionization Fraction *9*
2.2.2 The Formation of Molecular Hydrogen *13*
2.2.3 Cooling Functions *16*
2.3 Forming the First Stars *18*
2.3.1 Perturbations in the Early Universe *18*
2.3.2 Collapse of Perturbations in the Early Universe *21*
2.3.3 Cooling and the Jeans Instability *25*
2.3.4 Properties of the First Stars *31*
2.3.5 Remnants and Signatures of a Population III *34*
2.4 Primordial HII Regions *35*
2.5 What if Dark Matter is not Cold? *37*
2.6 Hints for Further Study *38*

3 The First Star Clusters and Galaxies *39*
3.1 Overview *39*
3.2 Subsequent Generations of Stars *39*
3.2.1 Second-Generation Population III Stars *40*
3.2.2 Population III Stars Forming in Self-Shielding Halos *41*

From First Light to Reionization. Massimo Stiavelli
Copyright © 2009 WILEY-VCH Verlag GmbH & Co. KGaA, Weinheim
ISBN: 978-3-527-40705-7

3.2.3	Late Population III Star Formation by Atomic-Hydrogen Cooling in Massive Halos 48	
3.2.4	Termination of the First Stars Phase 49	
3.3	Containing Gas in the Halos of Population III Stars 49	
3.3.1	Ionization Heating and Gas Temperature 49	
3.3.2	The Escape of Gas Heated by Ionization 53	
3.3.3	The Escape of Gas Following a Supernova Explosion 56	
3.3.4	Population II.5 58	
3.4	The First Star Clusters 59	
3.4.1	Clusters of Population III Stars and of Metal-Poor Stars 61	
3.4.2	The Origin of Globular Clusters 62	
3.5	The First Galaxies 63	
3.6	The First Active Galactic Nuclei 64	
3.6.1	Population III Black Holes 64	
3.6.2	Black-Hole Mergers 65	
3.6.3	The Highest-Redshift QSOs 66	
3.6.4	Direct Collapse to Black Holes 67	
3.7	Low-Metallicity HII Regions 67	
3.8	Numerical Techniques and Their Limitations 68	
3.8.1	Collisionless Dynamics 68	
3.8.2	Collisionless Dynamics: Particle-Mesh Codes 70	
3.8.3	Collisionless Dynamics: Treecodes 71	
3.8.4	Gas Dynamics 71	
3.8.5	Gas Dynamics: Smooth Particle Hydrodynamics 72	
3.8.6	Gas Dynamics: Eulerian Codes 72	
3.8.7	Improving Resolution Through Mesh Refinement 73	
3.8.8	Radiative Transfer 73	
3.9	Hints for Further Study 73	
4	**Cosmic Reionization** 75	
4.1	Overview 75	
4.2	The Properties of the Sources of Reionization 76	
4.2.1	The Surface Brightness of Reionization Sources 77	
4.2.2	Reionization in a Hydrogen-Only IGM 79	
4.2.3	Reionization in a Hydrogen–Helium IGM 80	
4.2.4	Results for a Homogeneous IGM 81	
4.2.5	Mean Metallicity at Reionization 83	
4.3	Adding Realism to the Calculations 85	
4.3.1	Escape of Ionizing Photons 85	
4.3.2	Clumpy IGM 88	
4.3.3	Two-Parameter Models 90	
4.4	Luminosity Function of Ionizing Sources 90	
4.4.1	Detecting Lyman α from Ionizing Sources 92	
4.5	Reionization by Population III Stars 95	
4.6	How Is the Intergalactic Medium Enriched? 96	

4.7	Reheating of the Intergalactic Medium	97
4.8	Keeping the Intergalactic Medium Ionized	98
4.9	Hints for Further Study	100

Part 2 Observational Techniques and their Results *101*

5 Studying the Epoch of Reionization of Hydrogen *103*
- 5.1 Overview *103*
- 5.2 Gunn–Peterson Troughs in Redshift 6 QSOs *104*
- 5.2.1 A Simple Gunn–Peterson Test *104*
- 5.2.2 The Gunn–Peterson Trough *106*
- 5.2.3 Lyman Series Lines *108*
- 5.2.4 Metal Lines *109*
- 5.2.5 HII Region Size Test *109*
- 5.2.6 Dark Gaps *110*
- 5.2.7 An Assessment of the Indication from QSOs Spectra *110*
- 5.3 Lyman α Sources as Diagnostics of Reionization *111*
- 5.3.1 Effect of a Finite Lyman α Line Width *111*
- 5.3.2 Intrinsic Properties of Lyman α Emitters *111*
- 5.3.3 Effect of a Local Ionized Bubble *114*
- 5.3.4 A Realistic Lyman α Escape Model *115*
- 5.3.5 Perspectives on Studying Reionization with Lyman α Sources *117*
- 5.3.6 Faint Lyman α Halos *119*
- 5.4 Neutral-Hydrogen Searches *121*
- 5.4.1 Other Applications of High-z 21-cm Observations *124*
- 5.5 Compton Optical Depth *126*
- 5.6 Lyman α Signature in the Diffuse Near-IR Background *127*
- 5.7 Hints for Further Study *128*

6 The First Galaxies and Quasars *129*
- 6.1 Overview *129*
- 6.2 The Lyman-Break Technique *129*
- 6.2.1 The Lyman Break as a Function of Redshift *130*
- 6.2.2 Synthetic Stellar Population Models *131*
- 6.2.3 Redshift 6 Dropout Galaxies *132*
- 6.2.4 Lyman-Break Galaxies at Redshift Greater than 6 *133*
- 6.3 The Lyman α Excess Technique *135*
- 6.4 The Balmer-Jump Technique *136*
- 6.4.1 An Old Galaxy at Low or High Redshift? *137*
- 6.5 Photometric Redshifts *139*
- 6.6 Samples of High-Redshift Galaxies *141*
- 6.6.1 Lyman-Break Galaxies at $z = 6$ *141*
- 6.6.2 Lyman-Break Galaxies at $z > 7$ *143*
- 6.6.3 Lyman α Emitters *144*
- 6.6.4 High-Redshift QSOs *146*
- 6.7 Fluctuations *147*

6.8	Direct Detection of the First Stars	148
6.9	Hints for Further Study	149

7 Deep Imaging and Spectroscopy Surveys 151

7.1	Overview	151
7.2	Field Choice for a Deep Imaging Survey	151
7.3	Observing Techniques for Deep Imaging Surveys	154
7.3.1	General Considerations	154
7.3.2	Dithering	154
7.3.3	Super Bias and Super Dark	155
7.3.4	Flat Fielding	156
7.3.5	Image Combination	157
7.4	Self-Calibration	160
7.5	Catalogs	161
7.5.1	Layout of an Automated Photometry Algorithm	162
7.5.2	Sextractor Photometry Tips	163
7.5.3	Simulations	164
7.6	Cosmic Variance	166
7.7	The Gravitational Telescope	166
7.8	Deep Spectroscopy	169
7.8.1	Spectroscopic Analysis Techniques	169
7.8.2	Slit Spectroscopy of Faint Targets	170
7.8.3	Slitless Spectroscopy	171
7.9	Hints for Further Study	171

8 The Reionization of Helium 173

8.1	Overview	173
8.2	Gunn–Peterson Troughs in QSOs	173
8.3	Constraints from the Temperature of the IGM	175
8.4	Change in Metal-Line Ratios	176
8.5	Change in HI Lyman α Forest	177
8.6	Reionizing Hydrogen First and Helium Later	177
8.7	A Limit on the Escape Fraction from Galaxies at $z \simeq 3$	178

9 Future Instrumentation 181

9.1	Overview	181
9.2	The James Webb Space Telescope	181
9.2.1	Historical Remarks	182
9.2.2	The JWST Science Requirements Document	183
9.2.3	Overview of JWST Instrumentation	184
9.3	Other Space-Based Instrumentation	185
9.3.1	The Wide-Field Planetary Camera 3	185
9.3.2	The Cosmic Origins Spectrograph	186
9.3.3	A Possible Future Large Telescope in Space	186
9.4	Large Ground-Based Telescopes	189
9.4.1	Ground vs. Space Comparison	189

9.4.2	Multiobject Spectroscopy	*191*
9.4.3	Very High Resolution Imaging	*191*
9.5	Observing 21-cm Radiation at High Redshift	*191*
9.5.1	Murchison Wide-Field Array	*192*
9.5.2	Low-Frequency Array	*192*
9.5.3	Square-Kilometer Array	*193*
9.5.4	A Radiotelescope on the Far Side of the Moon	*194*
9.6	Atacama Large Millimiter Array	*194*
9.7	Large Field of View Imaging	*194*
9.7.1	Large Synoptic Survey Telescope	*195*
9.7.2	Large Field of View Imaging from Space	*195*
9.8	Planck	*195*

A	**Overview of Physical Concepts**	*197*
A.1	Cosmological Quantities	*197*
A.2	Saha Equation	*198*
A.3	Polytropic Stars	*200*
A.4	Jeans Instability for a Two-Fluid System	*201*

Index *213*

1
Introduction

1.1
First Light and Reionization

The emergence of the first sources of light in the Universe and the subsequent reionization of hydrogen mark the end of the 'Dark Ages' in cosmic history, a period characterized by the absence of discrete sources of light. Despite its remote timeline, this epoch is currently under intense theoretical investigation and is beginning to be probed observationally.

There are various reasons why studying this epoch is important. The first reason is that the reionization of hydrogen is a global phase transition affecting the range of viable masses for galaxies. Before reionization small galaxies will be shielded by neutral hydrogen fromionizing UV radiation and therefore will be able to form moreeasily. After reionization and the establishment of a UV backgroundthe formation of very small galaxies is hampered [70, 77, 219] (these issues will be discussed in Chapters 3 and 5).

The second reason to study this epoch is that it makes it possible to probe the power spectrum of density fluctuations emerging from recombination at scales smaller than are accessible by current cosmic microwave background experiments (this can be done by different techniques, as discussed in Chapters 2 and 5).

Finally, in a Universe where structures grow hierarchically, the first sources of light act as seeds for the subsequent formation of larger objects. Thus, the third reason to study this period is that by doing so we may learn about processes relevant to the formation of the nuclei of present-day giant galaxies and perhaps on the connection between the growth of black holes and evolution of their host galaxies (these issues will be briefly touched upon in Chapter 3).

Once established, the importance of studying The End of the Dark Ages one might wonder how realistic it is to expect to be able to understand anything about something so remote and different from the conditions in the local Universe and with so few observational constraints when, e.g., star formation in our own galaxy is not yet fully understood. The practitioners of this field are confident that two facts make the formation of structures in the Dark Ages easier to study theoretically than similar processes occurring at other epochs:

From First Light to Reionization. Massimo Stiavelli
Copyright © 2009 WILEY-VCH Verlag GmbH & Co. KGaA, Weinheim
ISBN: 978-3-527-40705-7

Tab. 1.1 Adopted cosmological parameters.

H_0	70 km s^{-1} Mpc^{-1}	
Ω_m	0.26	including Ω_b
Ω_Λ	0.74	
Ω_b	0.045	
Y	0.26	He fraction by mass
$T_{CMB,0}$	2.728 K	present-day CMB temperature
σ_8	0.75	CDM power spectrum normalization

1. the formation of the first structures is directly linked to the growth of linear perturbations,

   and

2. these objects have a known – and extremely low – metallicity set by the end-product of the primordial nucleosynthesis.

Thus, our chances of success depend on our confidence in understanding the initial conditions and on the absence of many additional physical processes that are at play in the local Universe such as magnetic fields or torques generating angular momentum. For the same reasons we should expect our success at predicting the properties of following generations of objects to be less secure.

1.2
The Cosmological Framework

We adopt the standard cosmological framework of the Hot Big Bang theory and the values of the cosmological parameters derived from the WMAP concordance model [251] and shown in Table 1.1. In general, all numerical results will be derived for these cosmological parameters but we will on occasion discuss how different values of the cosmological parameters would impact the result.

The most critical of these cosmological assumptions is that dark matter is in the form of cold dark matter (CDM) as this guarantees a significant amplitude of the low-mass perturbations at very early epochs. Should dark matter be, e.g., in the form of warm dark matter, e.g., ~1 keV neutrinos, the low-mass power spectrum would be suppressed and the scenario for first light stars would be entirely different from the one discussed here (see brief discussion in Chapter 2).

1.3
Organization of this Book

This subject as well as the rest of cosmology or, more generally, physics is based on the interplay of theory and observations. One could imagine starting from an overview of the observational results and then moving on to their theoretical interpretation. I believe that such an approach would lead to many repetitions as several modern observational results can only be understood within a rather sophisticated theoretical framework. Thus, I have chosen the opposite approach of starting with theory. Students reading this book should not be led to believe that the logical structure that I have adopted has any resemblance to the actual historical development of the discipline. In reality, progress in this field (and elsewhere) is always made by multiple iterations of theory and observations. Each new observational results spawns several theoretical efforts, some of which are confirmed by follow up observations while some are falsified. The surviving theories of today are robust because they have undergone multiple cycles of 'natural' selection! The study of the Dark Ages has not yet undergone a full cycle of confrontation with observations. Thus, the level of reliability of any specific result and conclusion is not as high as that of the basic ideas and physical processes.

This book is structured into two parts. The first part is focused on all theoretical aspects of the formation of the first light sources and of the reionization of hydrogen. Whenever possible I have tried to use analytical arguments to highlight the physical processes at play, to derive approximate analytical results and to draw conclusions on their basis. However, in modern cosmology we have come to rely on numerical simulations for many detailed predictions or interpretation of data and I have described the results of simulation whenever necessary. As computing power continues to improve, increasingly more complex simulations with a more ambitious description of the physics and more elaborate algorithms become possible. Thus, one would expect that the results of present-day state-of-the-art simulations might be superseded or amended in the next few years. I hope that the basic physical insight derived from the analytical results will remain valid.

The second part of the book is dedicated to observational techniques and to present and future observations. We are now able to probe the evolution of the luminosity function of galaxies to redshift 6 and beyond. This allows us to begin probing the end of the Dark Ages. Future facilities will open up major new possibilities in this subject. Among them, I will devote particular attention to the James Webb Space Telescope, once dubbed 'The First Light Machine' and to the experiments, such as LOFAR, aimed at studying the reionization process through its signature in the neutral-hydrogen background.

The subject of this book is specialistic and it would be hard or impossible to propose exercises in the classical sense. However, I have added to many chapters a section containing suggestions for indepth study or calculations that are relevant to the subject matter and would increase the understanding of the subject by the reader. Often, there will be more than one way to complete these calculations but generally what is more interesting is the calculation itself and the related uncertainties than the numerical result.

The book also includes an appendix where some of the equations used in the main text are derived and discussed in more detail.

1.4
Key Observations in this Field

The main observables available to probe first light and reionization will be discussed in detail in the second part of the book, however, a quick overview may be useful to set the stage for the theory section.

Direct detection of Population III objects and of the first galaxies will be very challenging and it will be attempted by future deep imaging survey using techniques now in use at lower redshift, like the Lyman-break technique. Individual Population III stars could be detected most easily as supernovae and establishing the frequency of such objects is a crucial parameter affecting the feasibility of such surveys. Early objects may leave a signature in the backgrounds that could either be detected directly or through a fluctuation analysis (see Section 6.7). Detecting this signature may be simpler than detecting individual objects.

Polarization measurements with a microwave background experiment like WMAP enable us to constrain the Thompson optical depth (see (2.13) and Chapter 4) which is essentially a density-weighted number of free electrons along the line of sight. We can also probe directly the presence of neutral hydrogen by using the Gunn–Peterson trough [103] and the properties of Lyman α emitters. The Gunn–Peterson trough is essentially resonant Lyman α absorption of the UV continuum of distant objects for wavelengths below that of Lyman α. While diffuse neutral hydrogen present within some redshift interval will scatter the continuum, local hydrogen can scatter line emission and provide a somewhat complementary test to the Gunn–Peterson test. As we will see in Chapter 5, Gunn–Peterson trough constraints from distant quasars indicate that hydrogen is reionized at $z < 6$. Finally, a new promising area is that of 21-cm studies aiming at probing the distribution of neutral hydrogen at high redshift through detection of the 21-cm line emission or, in the most ambitious cases, of 21-cm line absorption over the cosmic microwave background (see Chapter 5).

Theory

2
The First Stars

2.1
Overview

This chapter is devoted to describing the processes starting from recombination and leading to the formation of the first stars and to the derivation of their properties. In this section we provide an overview of these processes while the following sections will fill in the details.

2.1.1
First Light

First light is a somewhat ill-defined concept. Suppose that some perturbation of a given rarity leads to the formation of the first discrete source of light in a given volume. Because of the nature of the CDM power spectrum we can expect that an even rarer perturbation will give rise to an earlier first source of light in a larger volume. A practical definition is then that first light is the first source of light in a given volume large enough that this source was unaffected by feedback from other earlier sources that might have formed elsewhere. A consequence of this definition is that in some volume a truly first light source may form after second-generation objects have already formed in some other volume.

Most researchers in the field believe that the first discrete sources of light in the Universe were stars. The first stars would consistitute the so-called Population III and enrich their surrounding medium to low – but well above primordial – metallicity. It is extremely likely that at least some of these first stars will leave a black hole as a remnant and such black holes could soon begin accreting and become mini-active galactic nuclei (hereafter AGN). Thus, it is only a matter of a few million years before mini-AGN begin to contribute to the ionizing background. Because of the inhomogeneities in the Universe, stars with primordial metallicity will continue to form in some locations after the first of these mini-AGN have turned on.

The CDM power spectrum and the dynamics of dark-matter fluctuations will play a major role in seeding the formation of the first objects by producing structures where baryons can cool at a faster rate than the diffuse medium. These dark-matter

From First Light to Reionization. Massimo Stiavelli
Copyright © 2009 WILEY-VCH Verlag GmbH & Co. KGaA, Weinheim
ISBN: 978-3-527-40705-7

structures are generally termed 'halos' for historical reasons and because dark matter is incapable of cooling and will in general occupy a more extended halo than the cooling gas.

Mini-AGN have also been proposed as the actual first light sources. In this case, they would form following the direct collapse to a black hole of a primordial density fluctuation. The most recent version of this idea has been proposed as a way to address the existence of very massive black holes powering the QSOs at redshift 6 discovered by the Sloan Digital Sky Survey [80] and relies on gaseous collapse in $10^8 M_\odot$ dark halos [19]. Black-hole formation by direct collapse would in this case be preceeded by the formation of Population III stars and their black-hole remnants. Thus, stars would in this case continue to be the first light sources but their remnants would not be seeding the formation of luminous quasars.

2.1.2
Forming the First Stars

In the following we will assume that the first light sources are indeed the first stars. In order to investigate the formation of the first population of stars we need to be familiar with the prevalent conditions in the early Universe. Our review of 'the first 6 minutes' will be only very cursory and the interested reader should find more detailed information in, e.g., the books by Peebles [206] and Peacock [205]. Within the paradigm of the standard Big Bang theory, the Universe begins its life in a hot, radiation-dominated phase. At early times the Universe undergoes an inflationary phase and the decoupling between radiation and many-particle species. The last of these decoupling events involves photons and neutrinos following the electron positron annihilation that increases the energy content of radiation by about a factor of two (24 per cent in temperature). Once the decoupling events are over, the Universe expands while cooling adiabatically. As the temperature decreases, the number of photons above the photoionization threshold of helium and hydrogen decreases and neutral species can form. For even lower temperatures a few molecules can form that will be important for allowing the cooling of baryons in the dark halos where the first stars are formed.

We will be primarily interested in molecular hydrogen but, in order to estimate its fractional density, we will need to derive the residual ionization fraction and the residual proton densities because electrons and protons act as catalyzers in the two main channels of molecular-hydrogen formation. Thus, we will start by deriving the residual ionization fraction. The four most common molecules after recombination and before the first stars are formed are, in order of decreasing abundance, neutral hydrogen, neutral helium, positively ionized hydrogen (i.e. protons), and molecular hydrogen. Of these, only neutral hydrogen and molecular hydrogen are effective coolants for the halos we are considering. We will see that each of them can cool halos capable of forming first stars but those halos will have very different masses, with possible consequences on the initial mass function of the first stars. Since first stars form out of unenriched material they start out with primordial metallicity – which is set by the primordial nucleosynthesis – and will begin enriching the inter stellar medium (ISM) and intergalactic medium (IGM) with metals through winds and, especially, supernova explosions.

A crucial and not fully answered question is how does the first star era terminate? Formation due to molecular-hydrogen cooling could be stopped by the destruction of molecular hydrogen but ultimately only chemical pollution by other stars or reionization will be able to stop the formation of first stars formed within halos cooling by atomic hydrogen. The relative importance of these effects will determine the star-formation history of the first stars and the possible existence of an intermediate stellar population with intermediate metallicity between Population II and Population III (the so-called Population II.5).

2.1.3
The Legacy of the First Stars

The first stars are part of the so-called Population III. However, because of inhomogeneities in metal enrichment it is possible that Population III stars will continue to form in the Universe long after an intermediate population or even Population II stars have started forming close to the sites of the first generation of stars.

We expect first stars to be hot, short lived, and very massive. Their temperature will make them very effective at ionizing hydrogen in their neighborhood as long as their ionizing radiation can escape the halos where they form. Ionizing radiation not escaping the halo will produce HII regions rather different from those that we are familiar with in the nearby Universe both because of their very low (likely primordial) metallicity and of their temperature. Due to their mass we expect the first stars to leave very massive black holes as remnants. These black holes may later act as seeds for AGN with important consequences on the formation and evolution of the host galaxies.

2.2
Before the First Stars

In this section we set the stage for studying the formation of the first stars. In particular, we derive the ionization fraction of hydrogen and the primordial abundance of molecular hydrogen, and describe the cooling function for a primordial chemical composition.

2.2.1
Recombination and Residual Ionization Fraction

In order to derive the residual ionization fraction after the recombination of hydrogen we can start at a redshift comfortably preceeding recombination of hydrogen but following the recombination of helium. The small number density of helium compared to hydrogen (about 8 per cent) and its much higher ionization energy (24.6 eV for He I against 13.6 eV) will ensure that the residual ionization fraction of helium can be neglected. At redshifts around 1500, physical conditions are such that hydrogen recombination and ionization can be considered in equilibrium. We can also ignore the influence of the energy released during recombination on the cosmic microwave background (hereafter CMB) temperature because of the large number of photons per baryon (~10^{10}). Thus, we can simply compute the ionized fraction of hydrogen x as a fraction of the decreasing CMB temperature using Saha's equation:

$$\frac{n_e n_p}{n_{HI}} = \frac{g_e g_p}{g_H} \frac{(2\pi m_e kT)^{3/2}}{(2\pi\hbar)^3} e^{-B_1/kT} \tag{2.1}$$

where n_e, n_p, and n_{HI} are, respectively, the number densities of electrons, protons, and neutral (atomic) hydrogen, $g_e = g_p = 2$, $g_H = 4$ their state degeneracy, m_e is the electron mass, k is Boltzmann's constant, and $B_1 = 13.6$ eV is the ionization energy of hydrogen. The total number density of protons is $n_H \equiv n_p + n_{HI} = 8.32 \times 10^{-6} \Omega_B (1+z)^3$ cm^{-3}, where h is the reduced Hubble constant, with $h = 0.7$ for our adopted value of H_0, and the ionization fraction is $x \equiv \frac{n_p}{n_H} = \frac{n_e}{n_H}$. Considering $T = 2.728(1+z)$ K, dividing (2.1) by n_H, and replacing whenever possible density ratios with ionization fractions x, we find:

$$\mathrm{Log}_{10} x^2 = 20.99 - \mathrm{Log}_{10}\left[\Omega_b h^2 (1+z)^{1.5}\right] - \frac{25152}{1+z} \tag{2.2}$$

In Figure 2.1 we plot this profile as a function of redshift (dotted line). For redshifts below 1000 the assumption of equilibrium is no longer valid and one has to solve for the ionization equilibrium taking into account the balance between ionization and recombination. Before doing so we need to derive a suitable equation by, e.g., following the approach by Peebles [207]. We start by writing the variation with time of the number density of electrons (in a comoving volume) as:

$$\frac{dn_e}{dt} = -\alpha_B n_e n_p + \beta_H n_{2s} \tag{2.3}$$

where α_B and β_H are, respectively, the (case B [196, 197]) recombination coefficient and the ionization coefficient and n_{2s} is the number density of atoms in the 2s state. Equation (2.3) essentially tells us that the variation in the number of electrons is the difference between the number of free electrons produced in ionization from the 2s (which is more accessible than the fundamental state because it requires only one quarter of the energy) and the number of electrons *used up* in recombination processes producing neutral atoms. We ignore recombinations to the ground state as appropriate in case B as they produce another ionizing photon. The term $\beta_H n_{2s}$ can be derived from the weighted sum of the ionization coefficients from all levels above the ground state, assuming that their populations are thermally distributed. In practice, one can relate α_B and β_H by considering that ionization and recombination processes tend to bring the ratio $\frac{n_e n_p}{n_{2s}}$ to its thermal equilibrium value given by the Saha equation, so that:

$$\beta_H = \alpha_B \frac{(2\pi m_e kT)^{3/2}}{(2\pi\hbar)^3} e^{-B_2/kT} \tag{2.4}$$

where $B_2 = 3.4$ eV is the binding energy of the 2s state of hydrogen, and m_e is the electron mass.

If the ground state and the 2s state were in thermal equilibrium one could compute n_{2s} from $n_{1s}e^{-(B_1-B_2)/kT}$. However, this is not the case. The ground state 1s can be populated by decay of the 2s state through a 2-photon process or from the 2p states by Lyman α emission. The Lyman α photons have enough energy to excite another atom until cosmological expansion brings them out of resonance. These factors have been captured by Peebles into a corrective coefficienct C_P given by:

$$C_P = \frac{1 + K_z \Lambda_{2s-1s} n_{1s}}{1 + K_z(\Lambda_{2s-1s} + \beta_H) n_{1s}} \tag{2.5}$$

where $K_z = \frac{\lambda_\alpha^3}{8\pi H(z)}$ is the cosmological factor redshifting Lyman α photons (with wavelength $\lambda_\alpha = 1215.67$ Å) out of resonance, $H(z)$ is the Hubble constant at redshift z, and $\Lambda_{2s-1s} = 8.22458$ s^{-1} is the relevant 2-photon decay rate.

Dividing now both sides of (2.3) by n_H, replacing the derivative with respect to time with one with respect to redshift, including the corrective coefficient C_P, equating n_{1s} to n_{HI}, and replacing where applicable density ratios by ionization fractions we find:

$$\frac{dx}{dz} = C_P \left[\alpha_B n_{HI} x^2 - \beta_H (1-x) e^{-(B_1-B_2)/kT} \right] \frac{dt}{dz} \tag{2.6}$$

The differential relation between time and redshift is given by:

$$\frac{dt}{dz} = -\frac{1}{H(z)(1+z)} \tag{2.7}$$

with

$$H(z) = H_0 \sqrt{\Omega_\gamma (1+z)^4 + \Omega_m (1+z)^3 + \Omega_R (1+z)^2 + \Omega_\Lambda} \tag{2.8}$$

where $\Omega_R \equiv 1 - \Omega_\gamma - \Omega_m - \Omega_\Lambda \simeq 0$ for our choice of cosmological parameters.

An equation similar to (2.6) but somewhat more general was derived by Jones and Wyse [124] who also discussed analytical approximations based on the Riccati form of this equation.

Following Peebles, we also adopt:

$$\alpha_B = 2.84 \times 10^{-13} \left(\frac{T}{10^4 \text{ K}}\right)^{-1/2} \text{ cm}^3 \text{ s}^{-1} \tag{2.9}$$

In order to investigate the effect of the corrective factor C_P we have first integrated (2.6) adopting $C_P = 1$. This is shown in Figure 2.1 as the long dashed line. Inclusion of the full expression from (2.5) for the factor C_P renders the equation *stiff*, requiring the use of a more sophisticated numerical integrator than the classical Runge–Kutta or predictor-correctors, e.g., the DVODE integrator [42]. The resulting curve is shown as the short dashed line in Figure 2.1. Clearly there is a major difference at high redshift just following recombination.

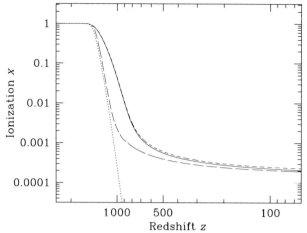

Fig. 2.1 Residual ionization as a function of redshift for different approximations. The Saha equilibrium approximation is only valid at high redshift and is shown by the dotted line. The results obtained by integration in the approximation of Peebles ([207], see also (2.6)) are shown by long and short dashed lines, respectively, with and without the corrective coefficient C_P. Finally, an effective model accounting also for the difference in temperature between gas and radiation is shown as the solid line.

Equation (2.6) relies on two major approximations: one is that all excited levels are thermally distributed and the other is the equality of the gas and radiation temperature. Both approximations were removed by Seager et al. [235] who considered a full set of equations including 300 levels. A simpler system providing essentially the same results was derived by Seager et al. [236]. Ignoring again the contribution of helium (included in Seager's analysis), we have:

$$\frac{dx}{dz} = C_P \left[\alpha_B n_{HI} x^2 - \beta_H (1-x) e^{-(B_1 - B_2)/kT_{gas}} \right] \frac{dt}{dz} \qquad (2.10)$$

$$\alpha_B = 1.14 \times 10^{-13} \frac{4.309 T_4^{-0.6166}}{1 + 0.6703 T_4^{0.53}} \text{ cm}^3 \text{ s}^{-1} \qquad (2.11)$$

and

$$\frac{dT_{gas}}{dz} = \frac{8\sigma_T a_R T_R^4}{3H(z)(1+z) m_e c} \frac{x}{1+x} (T_{gas} - T_R) + \frac{2T_{gas}}{1+z} \qquad (2.12)$$

Here, T_{gas} and T_R are, respectively, the gas and radiation temperatures, $T_4 \equiv (T/10^4 \text{ K})$, c is the speed of light, $\sigma_T = 6.652 \times 10^{-25} \text{ cm}^2$ is Thompson's cross section, and $a_R = \frac{\pi^2 k^4}{15 c^3 \hbar^3}$ is the radiation constant. The relation between β_H and α_B is still given by (2.4) with T_{gas} being used in place of the common temperature T.

These equations represent an effective approximation to the full system of Seager and collaborators [235]. The resulting residual ionization as a function of redshift is plotted as the solid line in Figure 2.1 and illustrates that the difference from the simple approach of Peebles [207] is modest. At a redshift of around 50 the residual

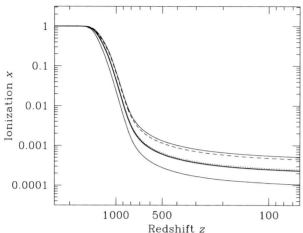

Fig. 2.2 Residual ionization for different values of the cosmological parameters are compared to the WMAP concordance model (thick solid line). The upper and lower thin lines differ from the concordance value only in the value of Ω_b taken to be 0.02 (upper) or 0.1 (lower). The dashed curve is for $\Omega = 1$ and $\Omega_b = 0.01$, while the dotted line is for $\Omega = 0.3$, $\Omega_\Lambda = 0$, and $\Omega_b = 0.0455$.

ionization is $x \simeq 2 \times 10^{-4}$. One can compute the redshift of the last scattering surface for this residual ionization profile by computing the Thompson optical length:

$$\tau = c\sigma_T \int_0^z \frac{n_e}{(1+z)H(z)} dz \tag{2.13}$$

and searching for the redshift where $\frac{d\tau}{dz}e^{-\tau}$ is maximum. By fitting a Gaussian to this (somewhat skewed) profile one derives $z = 1071 \pm 81$.

The residual ionization fraction depends on cosmology. We illustrate this in Figure 2.2 where we show for comparison residual ionization profiles for various cosmologies. The one for the adopted WMAP concordance model is shown as the thick solid line. The residual ionization at lower redshift ($z \approx 50$) can be approximated to a few per cent accuracy by:

$$x_{low-z} \simeq 1.38 \times 10^{-5} \frac{\Omega^{1/2}}{h\Omega_b} \tag{2.14}$$

2.2.2
The Formation of Molecular Hydrogen

Molecular hydrogen can form from the diffuse medium during the cooling phase following recombination and it can also form in individual collapsing halos. In this section, we will focus on the former to derive the cosmological abundance of molecular hydrogen. Our method will follow that of Lepp and Shull [137] even though our results will be numerically different because of the different input of atomic and cosmological parameters.

The two main channels of formation of molecular hydrogen at low densities are the H^- channel:

$$H + e^- = H^- + \gamma \tag{2.15}$$
$$H^- + H = H_2 + e^- \tag{2.16}$$

and the H_2^+ channel:

$$H + H^+ = H_2^+ + \gamma \tag{2.17}$$
$$H_2^+ + H = H_2 + H^+ \tag{2.18}$$

For both channels the first reaction is much faster than the second so that during the peak formation of H_2 essentially all H^- and H_2^+ are converted to H_2. The abundance derived from these reactions can be computed by writing down the corresponding rate equations:

$$\frac{dx_{H^-}}{dz} = -\frac{(c_{H3}x_e n_{HI} - c_{H4}x_{H^-} - c_{H5}x_{H^-} n_{HI} + c_{H16}x_e n_{H_2})}{H_0(1+z)E(z)} \tag{2.19}$$

$$\frac{dx_{H_2^+}}{dz} = -\frac{(c_{H8}x_p n_{HI} - c_{H9}x_{H_2^+} - c_{H10}x_{H_2^+} n_{HI} + c_{H15}x_{H_2} n_p)}{H_0(1+z)E(z)} \tag{2.20}$$

$$\frac{dx_{H_2}}{dz} = -\frac{(c_{H5}x_{H^-} n_{HI} - c_{H16}x_e n_{H2} + c_{H10}x_{H_2^+} n_{HI} - c_{H15}x_{H_2} n_p)}{H_0(1+z)E(z)} \tag{2.21}$$

where n_{HI} is the number density of neutral hydrogen and for a species A we use x_A to denote n_A/n_{gas}, where n_{gas} is the number density of all species in the gas. In these equations, we use the rates given by Galli and Palla [94]. Namely:

$$c_{H3} \simeq 1.4 \times 10^{-18} T_{gas}^{0.928} e^{-\frac{T_{gas}}{16200}} \tag{2.22}$$

$$c_{H4} \simeq 0.11 T_{rad}^{2.13} e^{-\frac{8823}{T_{rad}}} \tag{2.23}$$

$$c_{H5} \simeq 1.5 \times 10^{-9} \quad \text{for} \quad T_{gas} < 300 \text{ K}$$
$$4 \times 10^{-9} T_{gas}^{-0.17} \quad \text{otherwise} \tag{2.24}$$

$$c_{H8} \simeq 10^{-19.38 - 15.23 \log T_{gas} + 1.118 (\log T_{gas})^2 - 0.1269 (\log T_{gas})^3} \tag{2.25}$$

$$c_{H9} \simeq 1.63 \times 10^7 e^{-\frac{32400}{T_{rad}}} \tag{2.26}$$

$$c_{H10} \simeq 6.4 \times 10^{-10} \tag{2.27}$$

$$c_{H15} \simeq 3 \times 10^{-10} e^{-\frac{32050}{T_{gas}}} \quad \text{for} \quad T_{gas} < 1000 \text{ K}$$
$$1.5 \times 10^{-10} e^{-\frac{14000}{T_{gas}}} \quad \text{otherwise} \tag{2.28}$$

$$c_{H16} \simeq 2.7 \times 10^{-8} (T_{gas})^{-1.27} e^{-\frac{43000}{T_{gas}}} \tag{2.29}$$

Equations (2.19)–(2.21) can be integrated jointly with (2.10) and (2.12) to derive the final H_2 abundance. This is shown in Figure 2.3 where the solid line shows the total H_2 abundance while the dotted and dashed lines show, respectively, the abundances produced by the H^- and H_2^+ channels. H_2 can be produced also through other channels but the two considered here are the most important (see also [8]).

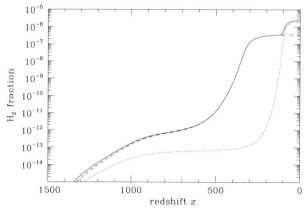

Fig. 2.3 Cosmological production of molecular hydrogen. The total fraction abundance is given by the solid line, while the dotted and dashed lines show, respectively, the abundances produced by the H^- and H_2^+ channels.

Tab. 2.1 Effective redshifts of formation of molecular hydrogen and relevant physical parameters.

	H^- channel	H_2^+ channel
z_{eff}	78	250
n_{HI} (cm^{-3})	0.098	3.15
T_{gas} (K)	108.6	616.5
x	2.23×10^{-4}	3.22×10^{-4}
R (cm^{-3} s^{-1})	1.08×10^{-16}	2.08×10^{-18}
t_H (s)	8.49×10^{14}	1.54×10^{14}

As noted by Lepp and Shull [137] one can obtain an estimate of the amount of H_2 produced by each channel assuming that production occurs in a burst with a duration of one local Hubble time and occurring at a specific redshift, different from the two channels. This is possible because at redshifts higher than the effective redshift, production is negligible because the molecules are efficiently destroyed. At redshifts much lower than the effective one, production is negligible because the density becomes too low. The first reaction in each channel is the critical one and one can estimate the quantity of H_2 produced by assuming that all H^- and all H_2^+ are converted to H_2. The abundance for each channel can be computed using the equation:

$$\frac{n_{HX}}{n_{HI}} = \frac{n_X}{n_{HI}} R_{HX} n_{HI}(z_{eff}) t_H(z_{eff}) \qquad (2.30)$$

where R_{HX} is the rate for the reaction being considered, z_{eff} is the effective redshift were the channel is most effective, t_H is the Hubble time at that redshift. The rate coefficient, neutral hydrogen densities (HI), electron (and proton) fractional density x, gas temperature, and the effective redshift are given in Table 2.1.

2.2.3
Cooling Functions

Cooling of a primordial gas made of essentially only atomic hydrogen and helium is efficient only for temperatures above 10^4 K. For lower temperatures collisions are almost never energetic enough to excite hydrogen or helium levels so as to enable radiative cooling. This is shown in Figure 2.4 where we show the cooling function for a primordial mix of atomic hydrogen and helium.

For this reason cooling by molecular hydrogen can be important even if its fractional number density is significantly lower than that of the atomic species. Indeed, for gas temperatures below $T_{gas} < 10^4$ K and above 10^2 K, the primary coolant is molecular hydrogen. The cooling rate per H_2 molecule can be given as a sum over all possible transitions of the energy difference between levels weighted by the fractional population of the states $f_{n,m}(T)$ and the spontaneous transition probability $P((n,m) \to (n',m'))$, namely:

$$\Lambda_{H_2}(T) = \sum_{n,m,n',m'} (E_{n,m} - E_{n',m'}) f_{n,m}(T) P((n,m) \to (n',m')) \tag{2.31}$$

In pratice, the calculation is more complex because one needs to take into account additional physics processes like reactive collisions changing the ratio of ortho

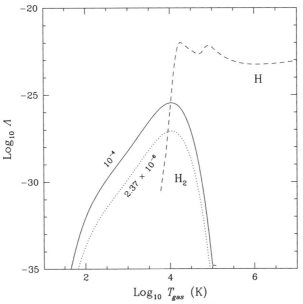

Fig. 2.4 Comparison of the atomic-hydrogen and helium cooling function (dashed line) with the one for molecular hydrogen for two values of the molecular-hydrogen fraction. The dotted line refers to the cosmological fraction 2.37×10^{-6}, while the solid line refers to the value of 10^{-4}. It is apparent that atomic cooling is much more efficient when the temperature is high enough.

and para hydrogen and collisional de-excitations. In Figure 2.4 we also show the molecular-hydrogen cooling function corresponding to the primordial fraction of molecular hydrogen derived previously and for the higher fraction of 10^{-4} that is relevant for the formation of structures. It is clear that cooling by molecular hydrogen is, in absolute terms, much less efficient than cooling by atomic species but is indeed the dominant cooling below 10^4 K.

In Figure 2.5 we show the molecular-hydrogen cooling function per molecule of H_2 derived by [94] for two values of the density. For hydrogen number densities not higher than 10^2 cm^{-3} it is possible to derive a simple analytical fit to these curves by noticing that in the range of temperatures of interest the cooling function of Figure 2.5 is roughly linear in a log–log plot. The relation:

$$\mathrm{Log}_{10} \Lambda(T_{gas}, n_{HI}) \simeq -31.6 + 3.4 \mathrm{Log}_{10}\left(\frac{T_{gas}}{100\ \mathrm{K}}\right) + \mathrm{Log}_{10}\left(\frac{n_{HI}}{10^{-4}}\right) \qquad (2.32)$$

is a very good approximation of the cooling function for temperatures between 120 and 6400 K. Under special circumstances cooling by other species such as, e.g., HD or LiH, can become important but in the following we will consider only the case of cooling by molecular and atomic hydrogen.

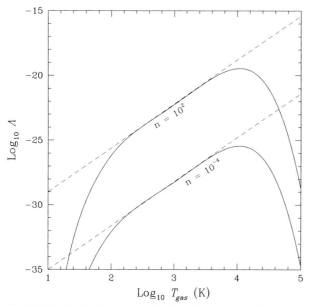

Fig. 2.5 Cooling function per molecular-hydrogen molecule derived by Galli and Palla for a hydrogen number density of 10^{-4} (lower solid line) and 10^2 (upper solid line) plotted as a function of gas temperature. The dashed lines represent the fits to these curves described in the text.

2.3
Forming the First Stars

In this section we describe the formation process of the first stars beginning from the growth of perturbation, and their collapse, and moving on to consider cooling of primordial halos, and the onset of their Jeans instability ending with formation of a Population III star.

2.3.1
Perturbations in the Early Universe

In this subsection we derive a number of results on the density contrast and mass function of perturbations in the early Universe. A significant simplification for our calculation is that at the high redshifts that we are interested in the Universe behaves as if $\Omega_m \simeq 1$ and $\Omega_\Lambda \simeq 0$. We can show this by deriving the evolution with redshift of Ω_m. The critical density $\varrho_c(z)$ is expressed in terms of $H(z)$, which is given by (2.8). This equation can be simplified considering that for our choice of cosmological parameters $\Omega_R = 0$ and that for $z \ll 1000\,\Omega_\gamma$, the contribution of radiation to the total mass density of the Universe, can be ignored. Thus, we find:

$$H(z) \simeq H_0 \sqrt{\Omega_{m,0}(1+z)^3 + \Omega_{\Lambda,0}} \tag{2.33}$$

where in this equation we have added the subscript $_0$ to quantities determined at the present time. Thus, for the critical density we find:

$$\varrho_c(z) = \frac{3H(z)^2}{8\pi G} \simeq \varrho_{c,0}(\Omega_{m,0}(1+z)^3 + \Omega_{\Lambda,0}) \tag{2.34}$$

The matter density in the universe ϱ_m at a redshift z is given by:

$$\varrho_m(z) = \Omega_{m,0}\varrho_{c,0}(1+z)^3 \tag{2.35}$$

The value of $\Omega_m(z)$ is simply the ratio of the densities in (2.34) and (2.35), namely:

$$\Omega_m(z) \simeq [1 + \Omega_{\Lambda,0}\Omega_{m,0}^{-1}(1+z)^{-3}]^{-1} \tag{2.36}$$

Thus, $\Omega_m(z) \simeq 1$ to better than 10% at $z \gtrsim 3$ and to better than 1% at $z \gtrsim 6.6$. It is easy to show that $\Omega_\Lambda(z)$ tends to 0 at the same redshifts. The fact that the Universe at $6 \lesssim z \lesssim 100$ behaves as a flat Einstein–de Sitter $\Omega_m = 1$ Universe simplifies all local calculations as it lets us ignore the presence of Ω_Λ. When computing quantities that are nonlocal, like the Hubble time t_H or the Thompson optical thickness τ (see (2.13)), we cannot ignore the effects of an integration over redshift and we must in general use the correct cosmology. For the Hubble time a simplification is still possible. The time as a function of redshift can be derived by integrating the derivative:

$$\frac{dt}{dz} = -\frac{1}{(1+z)H(z)} \tag{2.37}$$

2.3 Forming the First Stars

using for $H(z)$ the simplified expression of (2.33) with $\Omega_\Lambda = 0$, which is correct for $z > 6.6$. Integrating (2.37) from redshift ∞ to redshift z gives:

$$t_H(z) \simeq \frac{2}{3H_0\Omega_m^{1/2}}(1+z)^{-3/2} \simeq 3.34 \times 10^{15} \text{ s} \left(\frac{1+z}{31}\right)^{-3/2} \quad (2.38)$$

This quantity is approximate not only because we ignore Ω_Λ but also because at high redshift the contribution of radiation to the total matter density becomes important. In practice, the contribution of the high redshifts to the Hubble time is modest and the effective limit to the applicability of (2.38) is $z < 100$. Clearly, when computing the time from redshift 0 to redshift z we would instead need to integrate (2.37) without ignoring the term Ω_Λ.

Let us now consider the nonlinear evolution of a spherical overdensity for a flat Einstein–de Sitter Universe [206]. A shell of radius $r(t)$ containing a mass M moves according to the equation:

$$\ddot{r} = -\frac{GM}{r^2} \quad (2.39)$$

This is the same equation describing the evolution of the Universe scalelength and, in terms of an auxiliary variable η, it has the solutions:

$$r(t) = A(1 - \cos\eta) \quad (2.40)$$

and

$$t = B(\eta - \sin\eta) \quad (2.41)$$

where $A^3 = GMB^2$. The shell collapses to zero radius for $\eta = 2\pi$ when the time is $t_c = 2\pi B$. Let us now expand both (2.40) and (2.41) up to the 5th order in η [205]. We find:

$$r(t) = A\left(\frac{\eta^2}{2} - \frac{\eta^4}{24}\right) + O(\eta^6) \quad (2.42)$$

and

$$t = B\left(\frac{\eta^3}{6} - \frac{\eta^5}{120}\right) \quad (2.43)$$

The density contrast is:

$$\delta_c = \frac{3M}{4\pi r^3}(6\pi Gt^2) - 1 \quad (2.44)$$

where we have considered the mean density within the shell, $3M/4\pi r^3$, and the mean density in the background Einstein–de Sitter Universe, $(6\pi Gt^2)^{-1}$. Using the expression in (2.42) and (2.43) we find:

$$\delta_c = \frac{3\eta^2}{20} \quad (2.45)$$

We can now express B has $t_c/2\pi$ and derive η as a function of t from (2.43) (to third order) to derive:

$$\delta_c = \frac{3}{20}\left(\frac{12\pi t}{t_c}\right)^{2/3} \tag{2.46}$$

In the limit $t = t_c$, (2.46) gives us the value of the density contrast in linear theory corresponding to collapse and virialization, $\delta_c \simeq 1.68647$.

It is interesting to see what is the real density contrast produced by the nonlinear evolution. At the moment of maximum expansion $\eta = \pi$ (turnaround) we get from (2.40) $r = 2A$ and from the (2.41) $t = \pi B$. Replacing these (exact) values in (2.44) we find $\delta_c = 9\pi^2/16 \simeq 5.55$. Virialization leads to a decrease by a factor of 2 in radius leading to a density 8 times larger. At the same time the scale factor of the Universe has increased by a factor $2^{2/3}$ (from $t = \pi B$ to $t = t_c = 2\pi B$ and the mean density of the Universe has decreased with the cube of the scale factor, i.e. by a factor 4. Thus, the nonlinear density contrast of a virialized system is:

$$\xi = (9\pi^2/16) \times 8 \times 4 = 18\pi^2 \simeq 178 \tag{2.47}$$

Let us now review the standard analytical description of the mass function of dark-matter halos. Here, we follow the notation of Jenkins *et al.* [122]. For a power spectrum $P(k)$ and a linear growth factor of linear perturbations $b(z)$, normalized so that $b(z=0) = 1$ one can express the variance $\sigma^2(M, z)$ of the linear density field as follows:

$$\sigma^2(M, z) = \frac{b^2(z)}{2\pi^2}\int_0^\infty k^2 P(k) W^2(k; M)\,dk \tag{2.48}$$

where $W(k; M)$ is the Fourier space representation of the filter used to smooth out the density field at the mass scale M. Often for such a filter one adopts a real-space top-hat filter. The primordial power spectrum is generally taken to be linear in k but it is then modified by a suitable transfer function [29] taking into account that the original scale free power spectrum is modified by physical processes such as Silk damping [242] and free streaming. One can define the differential halo mass function as:

$$\frac{dn}{dM} = \frac{\varrho_0}{M}\frac{d\ln\sigma^{-1}}{dM}f(\sigma) \tag{2.49}$$

where ϱ_0 is the mean mass density of the Universe [221]. In this notation one obtains the familiar Press–Schechter approximation [214] by taking:

$$f_{PS}(\sigma) = \sqrt{\frac{2}{\pi}}\frac{\delta_c}{\sigma}\exp\left(-\frac{\delta_c^2}{2\sigma^2}\right) \tag{2.50}$$

where δ_c is the linearly extrapolated overdensity at the momentum of maximum compression (for a Universe with $\Omega_m = 1$, but in practice at $z > 6$ this is a good approximation even when $\Omega_m = 0.26$).

Detailed comparison with numerical simulations has suggested a different form for $f(\sigma)$ providing a better description of the halo mass function, namely the Sheth and Tormen (S–T) mass function [239]:

$$f_{ST}(\sigma) = A\sqrt{\frac{2a}{\pi}} \left[1 + \left(\frac{\sigma^2}{a\delta_c^2}\right)^p\right] \frac{\delta_c}{\sigma} \exp\left(-\frac{a\delta_c^2}{2\sigma^2}\right) \qquad (2.51)$$

where $A = 0.3222$, $a = 0.707$, and $p = 0.3$. While the Sheth–Tormen distribution is an improvement over the Press–Schechter one, it can still overpredict the abundance of rare objects by up to 50 per cent [221] and one should be aware of this uncertainty in what follows. We should also be cautious when applying the S–T distribution to rare objects at high-z because the distribution has not been fully validated in these conditions.

By tabulating numerically the function $\sigma(M)$ we can derive $\sigma(M,z) = \sigma(M)\delta(z)$ by rescaling $\sigma(M)$ with $\delta(z) \propto (1+z)^{-1}$. This would be exact in a Universe with $\Omega_m = 1$ but is an accurate approximation at $z > 6$ even for $\Omega_m = 0.26$ if we adopt an adjusted value for δ at zero redshift [47]. In ΛCDM and for redshift $(1+z) > \Omega_m^{-1/3}$ we can compute the growth of a perturbation with redshift $\delta(z)$ as:

$$\delta(z) = \frac{5\Omega_m}{2(1+z)} \left[\Omega_m^{4/7} - \Omega_\lambda + \left(1 + \frac{1}{2}\Omega_m\right)\left(1 + \frac{1}{70}\Omega_\lambda\right)\right]^{-1} \qquad (2.52)$$

If we now use (2.49) and (2.51) we can estimate the number of halos as a function of mass and redshift. This is shown in Figure 2.6 where we plot $dn/d\text{Log}_{10} M$ as a differential density per comoving cubic Mpc.

2.3.2
Collapse of Perturbations in the Early Universe

In order to begin forming stars the necessary condition is that the gas in a dark halo becomes Jeans unstable, i.e. that the Jeans mass in a halo becomes smaller than the total gas mass. The typical halos undergoing this process are the least massive ones for which this is possible as they will also be the most common.

The relevant cooling process (i.e. molecular versus atomic hydrogen) is determined by the virial temperature of the halos being considered. More massive halos will have higher virial temperature and will be capable of atomic-hydrogen cooling.

The mean density of a virialized dark halo at redshift z is given by:

$$\varrho_{vir} = \xi \Omega_M \varrho_0 (1+z)^3 \qquad (2.53)$$

where, following (2.47), $\xi \simeq 178$ is the ratio between the density of a virialized system and the background matter density of the Universe. An estimate of the radius of a virialized halo of mass M is:

$$R_{vir} = \left(\frac{3M}{4\pi \varrho_{vir}}\right)^{1/3} \qquad (2.54)$$

For a uniform system with constant density the half-mass radius is:

$$R_h = \frac{R_{vir}}{2^{1/3}} \tag{2.55}$$

Given the mass M and radius R_h, simple dimensional analysis provides the velocity scale. The gravitational potential energy per unit mass is $w = GM/R_h$. The kinetic energy per unit mass equals $(3/2)kT_{vir}/\bar{m}$, where k is Boltzmann's constant, $\bar{m} \simeq 1.242 m_p$ is the mean molecular weight for our adopted value of the helium mass fraction Y and T_{vir} is an estimate of the virial temperature. From the virial theorem we know that the kinetic energy per unit mass equals $w/2$ so that we can derive an expression of the temperature as a function of the mass M:

$$T_{vir} = \frac{1}{3} \frac{GM}{kR_h} \tag{2.56}$$

Here, we have essentially treated the virialized halo as a homogeneous density system even though this is not the case. Indeed, at least for massive halos the density is well approximated by a Navarro, Frenk and White mass distribution [182]. In practice, our expression for the virial temperature agrees within a few per cent with similar expressions derived in the literature [27, 271] and we will rely on (2.56) for a first exploration of the ability of gas in halos to cool. A more convenient way of expressing (2.56) is:

$$T_{vir} = 2525 \text{ K} \left(\frac{M}{10^6 M_\odot}\right)^{2/3} \left(\frac{1+z}{31}\right) \tag{2.57}$$

For the cooling function of Figure 2.4, a temperature of about 10^4 K separates the cases where atomic-hydrogen cooling is efficient from those where only molecular-hydrogen cooling is important. Equation (2.56) gives us the virial temperature as a function of mass. By selecting a virial temperature of 10^4 K we find that molecular-hydrogen cooling is important for halo masses below $\sim 10^7 M_\odot$.

Let us now estimate the range of masses and redshifts where objects within this mass range are able to cool efficiently. From the cooling function $\Lambda(T)$ we can derive a cooling timescale:

$$\tau_{cool} = \frac{3 n_{gas} k T_{vir}}{2\Lambda(T_{vir}) n_{H_2}} \simeq 1.63 \frac{k T_{vir}}{\Lambda(T_{vir}) f_{H_2}} \tag{2.58}$$

where we have expressed n_{H_2} in terms of the molecular fraction f_{H_2} as $n_H f_{H_2}$ and used the fact that $n_{gas}/n_H \simeq 1.087$ for our choice of Y. Equation (2.58) can be simplified using the approximation of (2.32) for the cooling function $\Lambda(T_{vir})$ to obtain:

$$\tau_{cool} = 3.64 \times 10^{10} \text{ s} \left(\frac{M}{10^6 M_\odot}\right)^{-1.6} \left(\frac{1+z}{31}\right)^{-5.4} f_{H_2}^{-1} \tag{2.59}$$

We can estimate the cooling efficiency from the ratio between the cooling timescale τ_{cool} and an evolutionary timescale taken to be the local Hubble time, $t_H(z)$. The latter choice is motivated by the fact that, in the idealized case a perturbation turning around at $t = t_H(z_{turn})$ recollapses at $t = 2t_H(z_{turn})$. We define cooling as efficient when $\tau_{cool}/t_H \leq 1$. In order to evaluate the efficiency we need to specify a value of the H_2 fraction. For a collapsing and cooling halo we can assume that the temperature will be to first order constant. By using the approximation for the formation of H_2 given in (2.30) and considering that the rate – which depends on the temperature – will also be to first order constant, we find that the H_2 fraction is to first order proportional to the density only. This yields an H_2 fraction of ~178 times the cosmic abundance of ~2×10^{-6}. This translates into a molecular-hydrogen fraction of ~3×10^{-4}. In reality, the H_2 fraction will depend on the virial temperature. A simple argument was proposed by Tegmark et al. [271] to derive an asymptotic value for the H_2 fraction, $f_{H_2}^{asymp}$ as the ratio of the molecular hydrogen production rate divided by the hydrogen recombination rate as this ratio gives the total production of H_2 occurring in the time necessary for all free electrons to recombine. With these assumptions and Tegmark's molecular reaction rates one finds $f_{H_2} \simeq 3.5 \times 10^{-4}(T_{vir}/1000 \text{ K})^{1.52}$. The same argument applied to Galli and Palla's molecular rates gives us:

$$f_{H_2}^{asymp} = 4.7 \times 10^{-4} \left(\frac{T_{vir}}{1000 \text{ K}}\right)^{1.54} \tag{2.60}$$

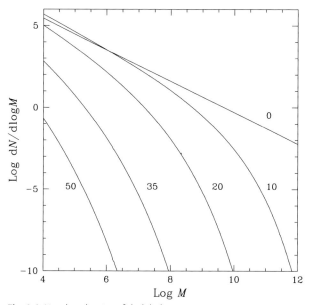

Fig. 2.6 Number density of dark halos per comoving Mpc as a function of redshift. Hierarchical clustering increases the mass scale of halos for decreasing redshift and at all redshifts higher-mass halos are rare. At redshifts above 30 the expected number varies very steeply with mass for masses in the range 10^6–$10^8 M_\odot$ because these masses are above the typical mass.

or, in a more convenient form, replacing (2.57) for the temperature T_{vir}:

$$f_{H_2}^{asymp} = 2.0 \times 10^{-3} \left(\frac{M}{10^6 M_\odot}\right)^{1.027} \left(\frac{1+z}{31}\right)^{1.54} \quad (2.61)$$

In practice, this expression tends to overestimate the final molecular fraction as the temperature decreases significantly during cooling so that the H_2 fraction computed from (2.61) is the total fraction of H_2 that would be formed if one suppressed cooling.

We can estimate the rarity of halos of a given mass at a given redshift by interpolating through the curves of Figure 2.6. In Figure 2.7 we show for a number of values of halo mass between $10^5 M_\odot$ and $10^8 M_\odot$ the redshift interval where the selected halo would be able to cool efficiently. The thin lines refer to the simple assumption of a constant molecular-hydrogen fraction, while the thick lines refer to a molecular-hydrogen fraction varying with the virial temperature according to (2.60). The plot shows that – if we focus on a volume no larger than one Gpc^3 – the first stars will be forming around $z = 50$ or slightly higher from dark halos of mass between 10^5 and $10^6 M_\odot$. Halos at the smallest mass end would not be able to cool in the case where the molecular-hydrogen fraction varies with the virial temperature as their virial temperatures are lower and their molecular-hydrogen fraction is then reduced.

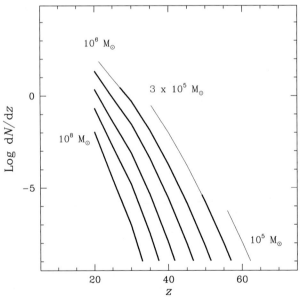

Fig. 2.7 As a function of redshift we plot the number density per unit redshift and per comoving Mpc^3 of halos that are able to cool. Left to right the curves refer to masses of $10^8 M_\odot$, $3 \times 10^7 M_\odot$, $10^7 M_\odot$, $3 \times 10^6 M_\odot$, $10^6 M_\odot$, $3 \times 10^5 M_\odot$, and $10^5 M_\odot$. The thin lines refer to the case of a constant molecular-hydrogen fraction, while the thick lines (partly overlapping the thin lines) refer to the case where the molecular-hydrogen fraction varies with the temperature.

By requiring $\tau_{cool}/t_H = 1$ we can derive the minimum H_2 fraction, $f_{H_2}^{min}$ needed for cooling. We estimate the Hubble time with (2.38) and use (2.59) for the cooling time to find:

$$f_{H_2}^{min} = 1.09 \times 10^{-5} \left(\frac{M}{10^6 M_\odot}\right)^{-1.6} \left(\frac{1+z}{31}\right)^{-3.9} \qquad (2.62)$$

We can now derive a minimum mass for a halo able to cool by noticing that $f_{H_2}^{min}$ is a decreasing function of the halo mass M, while $f_{H_2}^{asymp}$ as given by (2.61) is an increasing function of M. The physical interpretation for these trends is that more-massive halos are capable of forming a higher H_2 fraction but need a smaller fraction to cool efficiently. This must mean that there exists a value of the mass for which the two quantities are identical. At lower masses halos will not be able to form all the H_2 needed to cool. The minimum halo mass to cool, M_c can be derived by imposing the equality of (2.61) and (2.62) obtaining:

$$M_c = 1.36 \times 10^5 M_\odot \left(\frac{1+z}{31}\right)^{-2.071} \qquad (2.63)$$

The critical mass from (2.63) for redshift $z = 30$ is in excellent agreement with the results of numerical simulations [193]. It is worth noting that at the minimum mass cooling time is quite long and formation of a Population III star will show a significant delay.

2.3.3
Cooling and the Jeans Instability

The results of the previous section are based on an estimate of the molecular-hydrogen fraction in the collapsing halos. This estimate needs to be improved by considering the collapse of the system, the formation rate of molecular hydrogen and the actual cooling of the gas. A simple way to do this is to start from a virialized system with a cosmic fractional abundance of molecular hydrogen and a cosmic residual ionized fraction of hydrogen and to solve the equations of Section 2.2 for the case of a gas cloud collapsing at constant pressure (i.e. isobaric) as it cools. Tom Abel has written a program implementing this calculation (TOD [1]), and has made it available in the public domain. The only changes to the program that were necessary were the adoption of our cosmological parameters and our values of the residual ionized fraction and H_2 fraction derived in Section 2.2.

In Figure 2.8 we show how the gas number density, temperature and molecular hydrogen fraction vary with redshift for a dark halo with mass of $10^6 M_\odot$ forming at $z = 50$. It is remarkable that the asymptotic H_2 fraction is, in this case, very close to the value 3×10^{-4} estimated in the previous subsection.

In order to form a star the cooling gas must reach a temperature such that the gas Jeans mass becomes smaller than the total gas mass. This renders the cloud unstable to gravitational collapse and replaces the slow contraction taking place

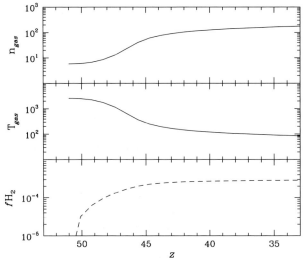

Fig. 2.8 Gas number density, temperature and molecular-hydrogen fraction as a function of redshift for a $10^6 M_\odot$ dark halo forming at $z = 50$. The calculation shows the evolution of this system computed using Tom Abel's public code and ignoring the onset of the Jeans' instability.

during cooling with a fast collapse. In principle, the collapse will continue until halted by another energy source such as shock heating generated by the infalling material hitting the expanding dense core at the center of the system or by the ignition of nuclear fusion.

A self-gravitating system is stable with respect to gravitational collapse if its mass is lower than its Jeans mass. The Jeans mass can be expressed as the mass within a sphere with a radius equal to one half the Jeans wavelength $\lambda_J = c_s \sqrt{\pi/G\varrho}$, where $c_s = dP/d\varrho$ is the adiabatic sound speed and ϱ the gas density. Thus, one finds:

$$M_J = \frac{\pi}{6} \varrho^{-1/2} \left(\frac{\gamma \pi k T}{G \tilde{m}} \right)^{3/2} \tag{2.64}$$

where $\gamma = 5/3$, T is the temperature, and $\tilde{m}$ is the mean mass per particle. In principle, one should compute the Jeans mass for the total, two-fluid, system of gas plus dark matter. Dark matter is unable to cool while gas cools and can become unstable. In practice, the dark-matter component plays a relatively neutral role in the instability (see the Appendix) and its density can be neglected after the gas has started to cool since the gas density increases by a factor of ~ 10 and becomes dominant. By replacing in (2.64), the baryonic density ϱ_b for the total density ϱ we find:

$$M_J = 8.4 \times 10^6 M_\odot \left(\frac{M}{10^6 M_\odot} \right) \left(\frac{\varrho_{b0}}{\varrho_b} \right)^{-1/2} \left(\frac{T}{T_0} \right)^{3/2} \tag{2.65}$$

where ϱ_b/ϱ_{b0} measures the increase in gas density during the cooling process and T/T_0 the decrease in temperature. In the simple case of isobaric cooling where $\varrho T = const$ we can obtain:

$$M_J = 8.4 \times 10^6 M_\odot \left(\frac{M}{10^6 M_\odot}\right)\left(\frac{T}{T_0}\right)^2 \qquad (2.66)$$

The baryonic mass is given by $M_b = \Omega_b M/\Omega_M \simeq 0.173M$.

When the Jeans mass exceeds the total gas mass of the halo, the gas is stable to gravitational collapse and continues to contract slowly as it cools. However, once the Jeans mass decreases below the total gas mass, the gas becomes unstable to gravitational collapse and begins to collapse. From (2.66) we see that in order to reduce the Jeans mass below the total baryonic mass during an isobaric collapse one needs a decrease in temperature by a factor of $\simeq 6.97$. We can determine a minimum mass limit for Jeans instability by considering the virial temperature of a halo from (2.57) and assuming that the minimum temperature achievable from H_2 cooling is 120 K (as for lower temperature the cooling function decreases rapidly, see, e.g., Figure 2.5). These two temperatures allow us to determine the maximum change in temperature for a halo of a given mass and redshift:

$$\frac{T_0}{T} = 21.04 \left(\frac{M}{10^6 M_\odot}\right)^{2/3}\left(\frac{1+z}{31}\right) \qquad (2.67)$$

Requiring now the maximum temperature change to exceed 6.97 gives us a minimum mass for Jeans collapse as a function of redshift:

$$M_{coll} = 1.905 \times 10^5 M_\odot \left(\frac{1+z}{31}\right)^{-3/2} \qquad (2.68)$$

Thus, in order to form a Population III star a halo has to have a mass satisfying both the cooling condition of (2.63) and the collapse condition of (2.68). The former is the most stringent condition for $z \lesssim 16.2$, while the latter is the strongest for $z \gtrsim 16.2$. A halo able to cool but not to collapse would have a Jeans mass always exceeding its total baryonic mass and would be coasting without ever being able to become gravitationally unstable. However, we should point out that to the level of the expected accuracy of our estimates the two conditions are essentially identical. During collapse the density will depart significantly from homogeneous and several of the approximations that we have used to derive (2.68) will break down. This is illustrated in Figure 2.12. A more accurate calculation would need to take into account the detailed density profile and be based on the Bonnor–Ebert [31,75] mass, representing the equivalent of the Jeans mass in the presence of an external pressure.

Using Abel's simple cooling code we can compute the Jeans mass as a function of redshift for any cooling halo. Figure 2.9 shows the decrease in Jeans mass for a $10^6 M_\odot$ halo collapsing at $z = 50$. To illustrate the case of cooling unable to lead to collapse we consider in Figure 2.10 the case of a halo of $1.7 \times 10^5 M_\odot$ at $z = 30$, i.e. a halo above the critical mass for cooling but below the mass required to have a Jeans collapse. As shown in the figure, Abel's cooling code predicts that this halo would ultimately Jeans collapse at $z < 6$. Clearly, after such a long delay the halo would have long lost its identity and would have merged with more-massive halos.

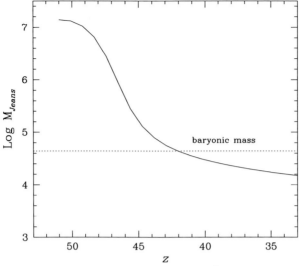

Fig. 2.9 Jeans mass calculated for the gas in a $10^6 M_\odot$ dark halo forming at $z = 50$. The dotted line represents the total gas mass in the halo. This halo become Jeans unstable at $z \simeq 45$.

Moreover, by redshift 6 reionization of hydrogen is completed, enabling heating of halos by the UV background and changing the rules of the game.

Running Abel's spherical collapse program for different values of the redshift and interpolating the results one finds that the redshift when gas in a $10^6 M_\odot$ halo becomes Jeans unstable, z_{Jeans}, is given with good approximation for $z \gtrsim 20$ by:

$$z_{Jeans} = 1.157 z_{form} - 11.73 \tag{2.69}$$

where z_{form} is the redshift of formation of the dark halo.

By adopting a minimum halo mass for the formation of Population III stars and assuming that only one star forms for each halo, we can derive the star formation rate of these objects, shown in Figure 2.11. There is evidence that halos close to the minimum mass can indeed only form one star but it is possible that more-massive halos can form more than one star [38]. In the figure we have ignored this possibility, which we will discuss in more detail in Section 3.4. Similarly, we have ignored the effect that a slow build up of the Lyman–Werner continuum has on the formation of these objects. We will discuss this in Section 3.2.4.

The precise upper limit to the mass of Population III is not well determined. The final mass of the star will depend on a number of factors like the total gas mass available and its angular momentum, the interaction with other nearby stars, and the characteristics of the accretion flow. Even focusing on the true first generation of stars and thereby ignoring feedback from others stars, the problem remains complex. Stars in accretion are luminous even before the ignition of nuclear fusion because of the energy released at the shock front where the accretion flow encounters the protostellar core [134]. Studies based on the dynamics of the accretion onto the protostellar core indicate the somewhat surprising possibility that

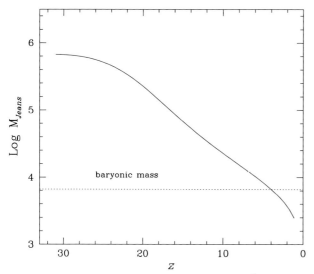

Fig. 2.10 Jeans mass calculated for the gas in a $1.7 \times 10^5 M_\odot$ dark halo forming at $z = 30$. The dotted line represents the total gas mass in the halo. This halo would become Jeans unstable only at $z < 6$.

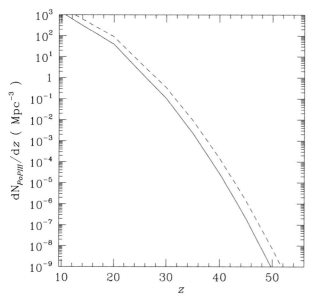

Fig. 2.11 Estimate of the star formation rate of Population III stars in units number of stars formed per comoving Mpc^3 and per unit redshift (used as a time unit). The solid line is computed applying a minimum halo mass of $10^6 M_\odot$, while the dashed line is for $6 \times 10^5 M_\odot$.

the final mass may be inversely proportional to the intensity of the accretion flow. When the flow is intense the star is bright and self-regulating. A mass of about $300 M_\odot$ would be reached at an accretion luminosity so high that the radiation pressure balances the gravitational force (see following subsection), thus terminating the accretion process [188]. A low accretion rate would enable formation of stars of mass up to $600 M_\odot$ [189]. However, even this is a simplification as the presence of angular momentum – acquired through torques – would turn the spherical accretion flow into an accretion disk weakening the argument based on Eddington luminosity. Thus, the expected mass of Population III stars is still relatively uncertain [308].

We do not even know for certain that a $\sim 10^6 M_\odot$ halo would form a single star as in some cases numerical simulations have indicated the existence of more than a single protostellar core [38] (see also Section 3.4). Thus, while the fact that Population III stars are generally very massive is probably a safe assumption, their initial mass function is essentially unknown at this time.

Our models so far assume that the collapsing halos are nonrotating and nonmagnetized. These assumptions are plausible for the first halos to collapse because fewer torques are available to generate angular momentum and these objects are the

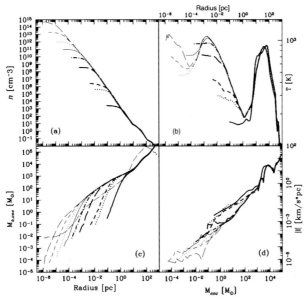

Fig. 2.12 Evolution of spherical average values for the baryon number density (upper left panel), baryon temperature (upper right panel), enclosed baryon mass as a function of radius (lower left panel), and baryon specific angular momentum (lower right panel). The thick solid line correspond to the onset of collapse at $z = 18.05$. The thick dotted, thick short dashed, thick long dashed and thick dot-dashed correspond to successive time intervals of 6 Myr, 0.19 Myr, 0.12 Myr, and 6810 yr later, respectively. The thin solid, thin dotted, thin short-dashed, thing long dashed correspond respectively to 782 yr, 151 yr, 58 yr and 14 yr later. The lighter dot-dashed line in the lower left panel is the Bonnor–Ebert critical mass calculated for the last step [193] (Reproduced by permission of the AAS).

first able to cool efficiently and therefore to collapse, making generation of magnetic fields by, e.g., dynamo amplification effect. The collapse of rotating and/or magnetized halos could lead to much more complex phenomena [149, 229].

2.3.4
Properties of the First Stars

We have seen that Population III stars are likely massive. A simple physical argument can give us an idea of their effective temperature [39]. The first assumption that we need to make is that the pressure-balancing gravity is essentially given by radiation pressure. This is reasonable given that we expect Population III stars to be very massive and therefore very luminous. The second – related – assumption that we need to make is that the star has a luminosity very close to its Eddington luminosity. The Eddington luminosity L_{Edd} is a luminosity such that the radiation pressure on the electrons equals the gravitational force on the protons. In such a case, neglecting the mass contribution of the electrons, charge neutrality – and the enormous electrostatic forces resulting from its violation – ensures that gravity and radiation pressure are balanced for the ionized gas as a whole. The force on the electrons F_e is given by:

$$F_e = \sigma_T \frac{L}{4\pi R^2 c} \qquad (2.70)$$

where σ_T is the Thomson cross section for the electron, L is the star luminosity and R its radius. The gravitational force on the protons is simply

$$F_p = -\frac{GMm_p}{R^2} \qquad (2.71)$$

where m_p is the proton mass. Equating the two forces from (2.70) and (2.71) gives us:

$$L_{Edd} = 4\pi GMm_p c/\sigma_T \simeq 3.3 \times 10^4 \left(\frac{M}{M_\odot}\right) L_\odot \qquad (2.72)$$

Assuming that the stellar luminosity is going to be very close to its Eddington luminosity is justified because the objects plausibly terminated accreting due to radiative feedback (see Section 2.3.3).

Dimensional analysis tells us that the central pressure – i.e. a force per unit area – must depend on the square of the mass and be inversely proportional to the fourth power of the radius, thus, for a star of mass M, radius R and, mass weighted, internal temperature T_i, we have:

$$P = 11.1 \frac{GM^2}{R^4} = \frac{1}{3}aT^4 \qquad (2.73)$$

Clearly, the constant 11.1 is not obtained by dimensional analysis but by describing the massive star as a $n = 3$ polytrope [51, 127] (see the Appendix). Using (2.73) to express T_i as a function of mass and radius we find:

$$T_i = 8.4 \times 10^7 \text{ K} \left(\frac{M}{100 M_\odot}\right)^{1/2} \left(\frac{R}{10 R_\odot}\right)^{-1} \qquad (2.74)$$

A Population III star begins its life by burning hydrogen using the p–p reaction. However, this reaction is not very efficient and in order to ignite it the star reaches internal conditions suitable to also ignite the 3α reaction burning helium into carbon. Thus, the star self-enriches very rapidly to the metallicity $Z \sim 10^{-9}$ needed to ignite the much more efficient CNO cycle [39]. For the sake of this analysis we can assume that the star is on the CNO cycle and describe its nuclear efficiency factor a function of metallicity Z, hydrogen fraction X, density ϱ and internal temperature T_i by adopting the following expression [39, 127]:

$$\varepsilon_{CNO} \simeq 1.3 \times 10^{12} ZX\varrho \left(\frac{T_i}{10^8 \text{ K}}\right)^8 \tag{2.75}$$

When averaged over the structure of the star, still taken to be a polytrope with $n = 3$, one finds that the average luminosity generation efficiency is $\langle\varepsilon\rangle \equiv L/M \simeq 0.05\varepsilon_{CNO}$, i.e. effectively only 5 per cent of the stellar mass takes part in the energy-generation process. Equating now the luminosity $\langle\varepsilon\rangle M$ to the Eddington luminosity L_{Edd} and deriving the temperature T_i as a function of metallicity Z, mass M and radius R, one finds:

$$T_i \simeq 1.9 \times 10^8 \text{ K} \left(\frac{Z}{10^{-9}}\right)^{-1/8} \left(\frac{M}{100 M_\odot}\right)^{-1/8} \left(\frac{R}{10 R_\odot}\right)^{3/8} \tag{2.76}$$

Finally, equating the temperature from (2.74) to that from (2.76) we find the mass–radius relation:

$$M \simeq 370 M_\odot \left(\frac{Z}{10^{-9}}\right)^{-1/5} \left(\frac{R}{10 R_\odot}\right)^{11/5} \tag{2.77}$$

The effective temperature of a star is the temperature that a blackbody of the same radius as the star would be required to have in order to maintain the same luminosity. This is a useful parameter considering that photons generated by nuclear reactions are heavily reprocessed and thermalized before leaving a star. We can find the effective temperature of a Population III star by equating its luminosity to the luminosity of a blackbody $4\pi R^2 \sigma T_{eff}^4$ and using the mass–radius relation of (2.77) to find:

$$T_{eff} \simeq 1.1 \times 10^5 \left(\frac{Z}{10^{-9}}\right)^{-1/20} \left(\frac{M}{100 M_\odot}\right)^{-1/40} \tag{2.78}$$

Thus, the effective temperature of a Population III star is essentially independent of its mass and is of the order of 10^5 K. This result is in good agreement with what is found in detailed numerical calculations [39, 281]. Stars of this temperature are very efficient at ionizing hydrogen and can also ionize helium. The properties of primordial HII regions will be discussed in more detail in Section 2.4.

Let us estimate the lifetime of one such star. The CNO cycle ultimately converts four protons into a helium nucleus (i.e. an α particle). The relative mass difference is $\eta \simeq 6.9 \times 10^{-3}$, which can be seen as the fraction of mass converted into energy.

The lifetime of the star $\tau_\star$ can then be determined from the ratio of the total energy reservoir of the star $\propto \eta M c^2$ divided by the Eddington luminosity (2.72). This gives:

$$\tau_\star = \frac{\eta \sigma_T c^2}{4\pi G m_p} \simeq 3 \times 10^6 \text{ yr} \qquad (2.79)$$

This quantity is independent of the mass of the star and it is a good approximation to the lifetime of Population III stars as determined by calculations of stellar evolution. Actual lifetimes are shorter than the above estimate by 20–30%.

Let us now derive the number of ionizing photons produced by Population III stars. We can start by assuming a blackbody spectrum with a temperature of $T_{eff} = 1.1 \times 10^5$ K (see (2.78)). The resulting rate of emission of ionizing photons per solar mass can be found as:

$$Q_0 = \frac{\int_{\nu_L}^{\infty} \frac{2\nu^2}{c^2} \frac{1}{\exp \frac{h\nu}{kT_{eff}} - 1} d\nu}{\int_{\nu_L}^{\infty} \frac{2h\nu^3}{c^2} \frac{1}{\exp \frac{h\nu}{kT_{eff}} - 1} d\nu} \times \frac{4\pi G M_\odot m_p c}{\sigma_T} \qquad (2.80)$$

where we have assumed that the star is at the Eddington luminosity. From (2.80) we obtain a value $Q_0 \simeq 2.6 \times 10^{48}$ ph s^{-1} $M_\odot^{-1}$. This value is about a factor of two

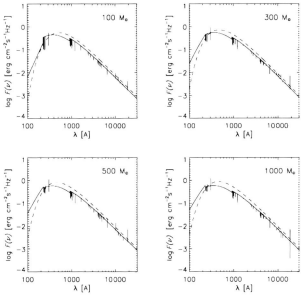

Fig. 2.13 Spectra of Population III stars of different mass from Bromm and coworkers [39]. In each panel a blackbody at 10^5 K is also shown for comparison (Reproduced by permission of the AAS).

Tab. 2.2 Rate of production of ionizing photons per unit mass for Population III stars.

Mass or mass range ($M_\odot$)	Q_0 ph s^{-1} $M_\odot^{-1}$	Notes and source
300	1.2×10^{48}	[39]
1000	1.6×10^{48}	[39]
300	1.25×10^{48}	ZAMS† [231]
300	1.34×10^{48}	life average [231]
1000	1.62×10^{48}	ZAMS† [231]
100–1000	1.6×10^{48}	[283]

† Zero-age main sequence

higher than what is obtained from more detailed calculations based on stellar atmospheres and stellar models, as shown in Table 2.2. This discrepancy is probably due to two factors. The main one is that stars do not really have blackbody spectra. In particular, they are underluminous compared to a blackbody at wavelengths just below the Lyman limit and are overluminous at shorter wavelengths [39] (see Figure 2.13). This excess of energy at shorter wavelengths compared to a blackbody of the same effective temperature reduces the number of ionizing photons as short-wavelength photons are more energetic. The second effect is that Population III stars have a luminosity slightly below the Eddington luminosity. It is interesting to see from Table 2.2 the very close agreement of the estimates from different groups.

2.3.5
Remnants and Signatures of a Population III

A fraction of the supermassive stars produced at zero metallicity may leave massive black holes as remnants. Whether or not a Population III star leaves a black hole remnant depends on the precise helium core mass at the end of its main sequence evolution and this is in turn related to its initial mass. In Figure 2.14 we show the initial vs. final mass function for a nonrotating Population III star. For masses between 140 and $260 M_\odot$ the life of the star is predicted to end as a pair-instability supernova. Pair-instability supernovae occur in very massive, radiation-dominated stars that are hot enough to trigger the formation of electron–positron pairs. This thermal loss reduces the degree of pressure support and causes the star to collapse and further increase its temperature leading to an explosive nuclear burning. A pair-instability supernova leaves no remnants [113]. At lower or higher masses a massive black hole remnant will be formed. For masses between 100 and $140 M_\odot$ the black-hole formation is preceded by pulsational pair instability, while for other masses one has direct collapse. It has been argued that under reasonable assumptions the total mass density of these black holes may be comparable to that of the supermassive black holes observed locally in the cores of giant galaxies [159]. This opens up the interesting possibility that the supermassive black holes are formed by merging of the population III remnants.

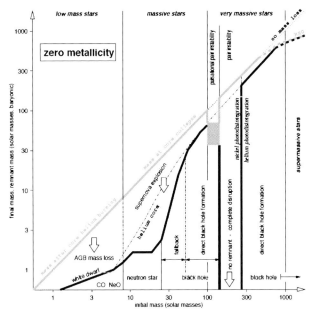

Fig. 2.14 Initial vs. final mass function for a nonrotating Population III star from Heger and Woosley [113] (Reproduced by permission of the AAS).

Population III stars and their supernovae are characterized by nucleosynthetic patterns different from those produced by ordinary stars [113, 187]. An important feature is the solar distribution of nuclei with an even nuclear charge coupled to a deficit in odd nuclear charge elements. Massive stars with nonprimordial composition can have stable burning phases following helium burning and these phases can produce elements with a neutron excess. These post-helium burning phases are thought to be absent in Population III stars. Massive Population III stars would also lack elements heavier than zinc due to the lack of s- or r-processes [113] but this second feature can be compensated if Population III stars of mass below $40M_\odot$ are also formed.

In principle, such patterns could be identified by studying low-metallicity objects at low redshift but possible contamination throughout cosmic history makes this measurement difficult to interpret. A possibility is to look for Population III elemental ratios in the Lyman α forest absorbers at high redshift.

2.4 Primordial HII Regions

Two direct consequences of the high effective temperature of zero-metallicity stars are their effectiveness in ionizing hydrogen (and helium) and their low optical-to-

UV fluxes. Both tend to make the direct detection of the stellar continuum much harder than the detection of the associated HII region.

Let us start out by considering an analytical estimate of the Lyman α luminosity of a primordial HII region. In case B all recombinations lead to either the 2s or the 2p state. The 2s state cannot directly decay to the ground-state 1s and has to undergo the much slower 2-photon process. The 2p state decays to the ground state by emitting a Lyman α photon. Thus, for a pure hydrogen composition and ignoring collisional effects, the number, $\dot{n}_\alpha$, of Lyman α photons emitted in the case B over the total number of ionizing photons, $\dot{n}_c$, is given by the ratio of the effective recombination coefficient to the 2p state, $\alpha^{\text{eff}}_{2\,2p}$, to the effective recombination coefficient α_B. For a temperature of 2×10^4 K we find from Osterbrock and Ferland [197] that $\alpha_B = 1.43 \times 10^{-13}$ cm^3 s^{-1} and that $\alpha^{\text{eff}}_{2\,2p} = 9.28 \times 10^{-14}$ cm^3 s^{-1}. To derive the latter number we have made use of the equality $\alpha_B = \alpha^{\text{eff}}_{2\,2s} + \alpha^{\text{eff}}_{2\,2p}$ with $\alpha^{\text{eff}}_{2\,2s} = 5.06 \times 10^{-14}$ [197]. For the case considered here the ratio $\alpha^{\text{eff}}_{2\,2p}/\alpha_B = \dot{n}_\alpha/\dot{n}_c = 0.646$ with only a modest temperature dependence. Indeed, in the low-density limit and within 5% for T between 5000 K and 20 000 K, roughly 2/3 of the ionizing photons produce a Lyman α photon. Collisions are important and can change this ratio. Considering now a hydrogen–helium mix and a non-negligible density we find from Panagia [201] that the ratio of Lyman α to total ionizing photons is:

$$\frac{\dot{n}_\alpha}{\dot{n}_c} \simeq \frac{1 + 1.35 \times 10^{-4} n_p}{1.5 + 1.35 \times 10^{-4} n_p} \tag{2.81}$$

where n_p is the proton number density. Thus, in the high-density regime the ratio $\dot{n}_\alpha/\dot{n}_c$ tends to unity. This is easily understandable as collisions can de-excite the long-living 2s state before it decays into two photons, thus relatively enhancing Lyman α decay from the 2p state.

Panagia et al. [203] report on calculations using Cloudy90 [82] of the properties of these HII regions (see also Figure 2.15). They find that the electron temperatures are in excess of 20 000 K and that 45% of the total luminosity is emitted by the HII region in the Lyman α line, resulting in a Lyman α equivalent width (EW) of 3000 Å [39]. The helium lines are also rather strong, with the intensity of HeII $\lambda1640$ comparable to that of Hβ [203, 282].

An interesting feature of these models is that the emission longward of Lyman α is dominated by a strong two-photon nebular continuum. The Hα/Hβ ratio for these models is 3.2. Both the red continuum and the high Hα/Hβ ratio could be naively (and incorrectly) interpreted as a consequence of dust extinction even though no dust is present in these systems. Because of the red nebular continuum the commonly assumed connection between metallicity and UV continuum [112] breaks down for Population III objects and a test to verify if a Lyman α source at high-z is a Population III object cannot be based on the UV continuum slope.

From the observational point of view, one will generally be unable to measure a zero metallicity but will usually be able to place an upper limit to it. When would such an upper limit be indicative that one is dealing with a Population III object?

A metallicity of $Z \simeq 10^{-3} Z_\odot$ is probably the smallest that can be effectively measured (see Figure 2.15). This value is lower than the minimum metallicity at reionization (see Section 4.2.5) but seems to be in agreement with the metallicity of the first galaxies [298]. A similar value is obtained by considering that the first supernova (SN) going off in a primordial cloud with a gas mass of $10^6 M_\odot$ will pollute it to a metallicity of $\sim 0.5 \times 10^{-3} Z_\odot$ [203]. Computing the number of SNae required to unbind the ISM in a primordial cloud provides an independent – and higher – value for the metallicity, $Z \simeq 10^{-2}$ [78].

2.5 What if Dark Matter is not Cold?

All the analysis so far is predicated on the assumption that dark matter in the Universe is cold and is characterized by a spectrum with power at very small scales. Recently, many groups have started considering the possibility that dark matter is warm [28], i.e. made of particles with lower mass and thus higher velocities that would introduce a cutoff at higher mass compared to cold dark matter (CDM). Depending on the exact mass of WDM one would suppress the perturbation spectrum at the masses of interest for the formation of Population III stars. In such a WDM Universe, Population III star formation would be suppressed and the first stars would be formed at much later redshift [192]. The basic arguments on the formation and properties of Population III stars made in Section 2.3 would still

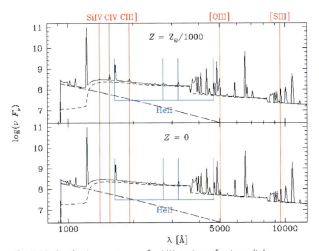

Fig. 2.15 Synthetic spectrum of a HII region of primordial metallicity (bottom) compared to one with a metallicity 10^{-3} the solar value (top). The main hydrogen and helium lines are identified in the figure [203]. The solid lines given the overall spectrum, the long dashed lines give the stellar SED, the short dashed lines the contribution of the nebular continuum.

be applicable but these objects would begin forming at lower redshift out of more massive halos (see (2.63)).

2.6
Hints for Further Study

- Solve for the residual ionization fraction also including helium and using the Seager et al. approximation [236]. A solver for stiff differential equations will be needed, e.g., DVODE [42].
- Using for the cooling function at $T = 100$ K the approximation $\text{Log}_{10}\Lambda = -29.51 + \text{Log}_{10}(Z/10^{-3}Z_\odot)$ determine as a function of redshift the minimum metallicity enabling a halo with this gas temperature to collapse.
- Determine for which effective blackbody temperature a Population III start would have the same ionizing photons output as that obtained from model atmospheres.
- Determine the minimum mass able to collapse for redshift $z = 20, 30, 40$ using Abel's code TOD [1] and compare results with the analytical approximations derived in this chapter.

3
The First Star Clusters and Galaxies

3.1
Overview

The formation of the first associations of stars is a much more complex problem than the formation of individual Population III stars and it is much harder to make statements on this problem without relying heavily on the results of numerical simulations. This chapter addresses separately the formation of the second generation of stars, and the formation of the first star clusters, the first galaxies and the first active galactic nuclei. The attempt is to describe a few basic ideas in each area and highlight possible directions of progress in the coming years.

3.2
Subsequent Generations of Stars

A first-generation Population III star can be defined as a Population III star that forms in isolation, i.e. without being affected by radiative, mechanical or chemical feedback from other stars. These objects have been referred to in the recent literature as Population III.1 objects [195]. However, it is possible that a Population III star will begin to form in the immediate vicinity of another one out of chemically pristine material, possibly because of multiplicity of accretion cores in the same halo or in a neighboring halo. Such objects would represent second-generation Population III stars, or Population III.2. Stars forming in the vicinity of other structures could also be subject to torques and the acquired angular momentum is also likely to affect the mass function of these objects. After a Lyman–Werner background is established and in halos not polluted by chemical elements one will be able to form additional generations of Population III stars either through molecular-hydrogen cooling in self-shielding halos or through atomic-hydrogen cooling in more massive halos. Generally radiative feedback from previous generations of stars can be positive or negative. A positive feedback is caused, e.g., by partial ionization of gas, as free electrons and protons are catalyzers for the formation of molecular hydrogen (see (2.15) and (2.17)). Conversely, ionization and destruction of H_2 molecules will generally suppress further star formation in a halo.

From First Light to Reionization. Massimo Stiavelli
Copyright © 2009 WILEY-VCH Verlag GmbH & Co. KGaA, Weinheim
ISBN: 978-3-527-40705-7

Recently, it has also been pointed out that Population III stars forming after the first Population III stars have produced supernovae may be affected by cosmic rays generated by these explosions. The extra ionization due to cosmic rays would favor the formation of additional molecular hydrogen and would enable the formation of stars of lower mass [254].

Halos may also be polluted by chemical elements produced by Population III stars. Such halos will then be incapable of forming true Population III stars but might have a metallicity so low that the stars being formed share some of the properties of Population III objects and are generally referred to as Population II.5. In the following we will review each of these types of stars and their associations.

3.2.1
Second-Generation Population III Stars

Second-generation Population III stars form in the vicinity of another star and will be subject to radiative feedback from the first star. This will affect the ability of the halo to form stars as well as the mass of the stars formed. Let us consider first a low-mass halo in the vicinity of a Population III star. Molecular hydrogen can be dissociated by nonionizing radiation through the Solomon process involving 76 Lyman–Werner resonances in the energy range 11.18–13.6 eV (with mean energy of 12.39 eV) [1, 108, 255]. This implies that molecular hydrogen cannot be shielded by atomic hydrogen and that LW photons will travel further than ionizing photons. Accordingly, one can expect that a Population III star in the vicinity of a low-mass halo will be able to destroy all primordial molecular hydrogen and perhaps partially ionize hydrogen. We have considered in Figure 3.1 the case of a halo of mass 2.5×10^5 mass collapsing at $z = 40$. Such a halo is able to cool and collapse slowly, achieving a Jeans collapse only at $z \simeq 20$. However, in the vicinity of a Population III star destroying all molecular hydrogen but introducing a partial ionization at the level of $x = 0.01$ this halo will cool and collapse more efficiently (solid line in the figure) because the additional free electrons can more effectively catalyze the formation of new molecular hydrogen. This is an example of the type of positive feedback produced by Population III stars.

We have seen in Section 2.3.3 that the mass of a star in a fast accretion flow is self-limiting due to radiation pressure. It is natural to conjecture that a second-generation star forming in the vicinity of the first would experience a more intense radiation field (up to a factor of two for a star in the same halo) and could be expected to have a lower mass, possibly much lower than that of a first-generation star. Similarly, angular momentum could cause fragmentation of the gas cloud and lead to the formation of a multiple system of lower-mass objects.

If the formation of the second-generation star takes place in a halo already photoionized by previous stars, accretion rates are going to be further decreased and the whole formation process may follow a very different path, e.g., through an accretion disk. Simulations by O'Shea and coworkers [191] suggest that these objects will have a mass in the range $5-20 M_\odot$ to be compared to the mass range $80-200 M_\odot$ predicted by the same authors for first-generation stars.

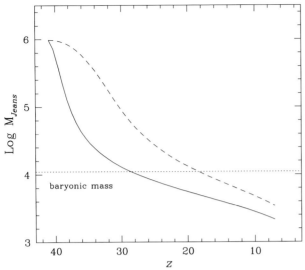

Fig. 3.1 Evolution with redshift of the Jeans mass of a halo with $M = 2.5 \times 10^5 M_\odot$ collapsing at $z = 40$. The dashed line represents the evolution of an isolated halo. The solid line shows the evolution of a halo partially ionized by a neighboring Population III star. The preionized halo is able to cool and collapse more efficiently.

3.2.2
Population III Stars Forming in Self-Shielding Halos

Let us start by considering whether primordial molecular hydrogen can effectively shield halos from the ultraviolet radiation of the first Population III stars. The Solomon process for dissociating molecular hydrogen with radiation in the Lyman–Werner bands is sensitive to the molecular-hydrogen column density of the system under consideration. For low values of the column density, absorption in the Lyman–Werner line wings is unlikely and only resonant photons will be able to dissociate H_2. As the column density increases more and more photons will be capable of dissociating H_2 [99]. At the redshift we are interested in ($z \gtrsim 10$), few massive structures have collapsed and the Universe is quasihomogeneous, so that the relevant column density is set by the distance over which radiation in the LW bands is redshifted out of those bands. Requiring that radiation at the bands center ($\varepsilon_c = 12.39$ eV) is redshifted below $\varepsilon_l = 11.18$ eV gives us a scalelength that combined with the molecular-hydrogen number density gives the molecular-hydrogen column density. Over the same distance photons from the upper half of the band will redshift into the lower half. Photons redshifting out of the bands are not compensated by the redshifting of more energetic photons into the LW bands until the Universe reionizes since those photons with energy above 13.6 eV are absorbed very effectively by neutral hydrogen. The probability of absorption of a LW photon depends on the column density of molecular hydrogen. We start by computing the column density of H_2 for a homogeneous expanding Universe over the physical distance corresponding to the redshift interval from the center to the edge of the LW band, namely:

$$D_c = \left[\frac{\varepsilon_c}{\varepsilon_l} - 1\right] \frac{c}{H(z)} \simeq 4.66 \text{ Mpc} \left(\frac{31}{1+z}\right) \frac{1}{\sqrt{1+\Omega_m z}} \quad (3.1)$$

where we have approximated $H(z) \simeq 70(1+z)\sqrt{1+\Omega_m z}$ km s^{-1} Mpc^{-1}, which is valid for a matter-dominated Universe [205] (also compare to (2.8) for $\Omega_A = \Omega_\gamma = 0$) and is good to better than 10% for $z \geq 10$. Recognizing that $n_{H_2} \simeq 2.4 \times 10^{-6} n_H$ and that $n_H \simeq 5.5 \times 10^{-3}[(1+z)/31]^3$ cm^{-3} we find that the column density of molecular hydrogen is:

$$N_c = n_{H_2} D_c \simeq 2 \times 10^{17} \text{ cm}^{-2} \left(\frac{1+z}{31}\right)^2 \sqrt{1+7.8\left(\frac{z}{30}\right)} \quad (3.2)$$

The column density as a function of z, in agreement with the expression above but derived without approximation for $H(z)$, is shown in Figure 3.2.

Unfortunately, we cannot apply the molecular-hydrogen column density of (3.2) directly to obtain the fraction of absorbed LW photons using, e.g., the tabulated values by Glover and Brand [99]. The reason for this is that the redshiftings of photons cannot be ignored. Photons being emitted far from a resonant line would have a low probability of being absorbed in the absence of redshift. However, as they redshift across the LW band they will hit local line centers and will be absorbed much more efficiently. We can estimate the cumulative effect of redshift statistically. Let us consider a portion of the gas column with density 10^{14} cm^{-2}. This portion absorbes a fraction 10^{-3} of the LW photons [99]. The total column of (3.2) has 2×10^3 of such portions for a cumulative relative absorption $W_{tot} \simeq 2$. This is equivalent to adopting an effective optical depth of LW photons $\tau_{LW} \simeq 2$. In practice, the column density available to absorb photons with energy close to 13.6 eV is twice as long as that of (3.2) since this was computed from the band center. On the other hand photons close to 11.18 eV will be redshifted out of the LW bands for much shorter columns. Let us now divide the LW band in 4000 wavelengths bins with equal logarithmic spacing. By construction, redshift will span one such bin when traveling the distance equivalent to a 10^{14} cm^{-2} molecular-hydrogen column. The cumulative relative absorption width W_{tot} is thus given by:

$$W_{tot} = 10^{-3} \frac{\sum_{i=1}^{4000} i}{4000} = 10^{-3} \frac{4000 \times 4001}{2} \frac{1}{4000} \simeq 2 \quad (3.3)$$

Using the same formalism of Glover and Brand we can compute the effective absorption W_{eff} from W_{tot} as 82%. Approximating a Population III SED with a blackbody with effective temperature $T_{eff} \simeq 1.1 \times 10^5$ K (see (2.78)) we find that the LW photons represent 9.1% of the ionizing photons. Combined with the absorption of 82% we have that number of LW photons that can be absorbed before being redshifted is a fraction 7.5% of the ionizing photons [278]. A detailed stellar atmosphere would give us similar numbers, e.g., for a 300$M_\odot$ star the LW photons are

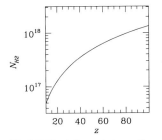
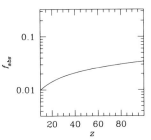

Fig. 3.2 The left panel shows the molecular-hydrogen column density at the distance at which half the emitted LW photons have redshifted out of band. This is the distance within which LW photons need to be absorbed in order to dissociate molecular hydrogen in the nonexpanding case. Note that these photons are not replaced because ionizing photons are efficiently absorbed by neutral hydrogen before they can redshift into the LW bands. The right panel shows the absorption fraction of LW photons ignoring the expansion of the Universe that, as we show in the text, brings the absorbed fraction close to unity.

8.5% of the ionizing photons [231] and the fraction that can be absorbed would go down to 7%. A more precise estimate would requiring modeling in detail the radiative transfer of LW radiation [6].

Let us now turn determine how effective a Population III star is at dissociating molecular hydrogen. The condition for dissociating intergalactic molecular hydrogen can be derived starting from the star-formation rate of Figure 2.11. Indeed, integrating the star-formation rate one derives the cumulative number density of Population III per comoving Mpc^3 as a function of redshift. In principle, one should compute this quantity by considering explicitly the minimum mass of a halo capable of forming a Population III star as given by (2.63) and (2.68). Here, for simplicity we will simply assume two values of the minimum mass for halos capable of forming Population III stars: $10^6 M_\odot$ (solid line) or $6 \times 10^5 M_\odot$ (dashed line) as this allows us to bypass the issue of the delay in the formation of a Population III star for masses close to the threshold. In Figure 3.3 we show the cumulative number of Population III stars for the two values of the minimum mass. The emission of LW photons can be estimated as follows. As we have seen previously in the blackbody approximation, the number of photons in the LW bands is about 7.5% that of the ionizing photons. Assuming a typical stellar mass of $300 M_\odot$ and the rate of emission of ionizing photons from Table 2.2 we find that this object will emit an effective number of 2.5×10^{63} LW photons over its lifetime. Here, by effective we mean that this is the number of emitted photons that can be absorbed by molecular hydrogen before being redshifted off band. The probability of a LW that is absorbed to actually photodissociate molecular hydrogen is only about 15% [99, 241] so this implies that a Population III stars will destroy 3.7×10^{62} H$_2$ molecules over its lifetime. Since the density of hydrogen is 5.4×10^{66} Mpc^{-3} and the primordial H$_2$ fraction is 2.4×10^{-6} this implies that a Population III star will destroy H$_2$ over a comoving volume of about 28 Mpc^3. Thus, all primordial molecular hydrogen will be destroyed when the cumulative density of Population III stars reaches ~ 0.04 Mpc^{-3}. Based on our estimate in Figure 3.3 this occurs at redshift $z \simeq 30$ [150, 278]. Note also that the total number of molecular-hydrogen molecules

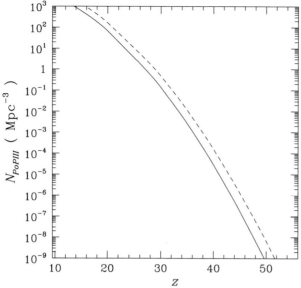

Fig. 3.3 Cumulative number density of Population III stars per comoving Mpc^3 as a function of redshift. The solid line is computed applying a minimum halo mass of $10^6 M_\odot$, while the dashed line is for $6 \times 10^5 M_\odot$.

forming in a $10^6 M_\odot$ halo is 10^{59}, which is 3 orders of magnitude lower than the number of H_2 molecules destroyed by a single Population III star. Thus, at $z < 30$ Population III star formation by H_2 cooling is strongly suppressed in any halo incapable of self-shielding. This results is in broad agreement with the result of radiative transfer simulations [123].

The background flux density required to photodissociate all H_2 can be derived easily from the density of LW photons derived above, $n_{LW} \simeq 10^{62} \gamma$ comoving Mpc^{-3}. Multiplying the number density by the photon energy and dividing by the frequency leaves one with Planck's constant. Thus, we have:

$$J_{LW} \simeq \frac{n_\gamma hc}{4\pi}(1+z)^3 \simeq 1.58 \times 10^{-24} \text{ erg s}^{-1} \text{ cm}^{-2} \text{ Hz}^{-1} \text{ sr}^{-1} \cdot \left(\frac{n_\gamma}{10^{62} \text{ Mpc}^{-3}}\right)\left(\frac{1+z}{31}\right)^3 \quad (3.4)$$

where the $(1+z)^3$ is needed to convert the photon density from comoving to physical units. The flux density of the background is commonly given in $J_{21} = 10^{21} J_{LW}$ so that we have $J_{21} = 1.58 \times 10^{-3}$ as the background to photodissociate all primordial H_2.

Let us now estimate the level of LW background needed for suppressing Population III star formation. Inspired by the criterion of collapse requiring the cooling time to be lower than the local Hubble time, one might in principle obtain the photodissociation condition by requiring that the asymptotic fraction of H_2 be destroyed within a local Hubble time. Unfortunately, the photodissociation cross section due to LW photons is not large and the above assumption would be equivalent to assuming that the photon mean free path is smaller than the size of halos, which is not the case. This incorrect assumption would lead one to overestimate the effectiveness of the LW background. A better method to determine the suppression of molecular-hydrogen cooling consists in requiring that the H_2 formation rate and its photodissociation rate are equal so as to derive the equilibrium molecular-hydrogen fraction [148] and then to compare this fraction with that needed for cooling from (2.62). We can write the timescale for photodissocation as follows:

$$\tau_{diss} = \frac{1.0}{4\pi k_{27} J_{LW}} \tag{3.5}$$

where $k_{27} = 1.1 \times 10^8$ is the H_2 photodissociation rate [1]. Replacing the coefficient and introducing $J_{21} = J_{LW} 10^{21}$ one finds:

$$\tau_{diss} = \frac{7.16 \times 10^{11}}{J_{21}} \text{ s} \tag{3.6}$$

The formation timescale for H_2 in a collapsing halo can be computed directly form the rate of the H^- formation channel of (2.15) that is dominant at lower redshifts, ignoring dissociations and other channels:

$$\tau_{form} = \frac{n_{H_2}}{c_{H3} n_{HI} n_e} \simeq \frac{f_{H_2}}{c_{H3} n_e} \tag{3.7}$$

where f_{H_2} is the desired H_2 fraction, c_{H3} is the reaction rate given by (2.22), and n_e is the electron density. The rate c_{H3} depends on the temperature but for a virialized halo we can express it more conveniently in terms of the mass and redshift by replacing for T the expression of (2.57) to find:

$$c_{H3} = 2 \times 10^{-15} \left(\frac{M}{10^6 M_\odot} \right)^{0.6187} \left(\frac{1+z}{31} \right)^{0.928} \tag{3.8}$$

From (2.14) (see also Figure 2.1) we derive the ionized fraction at later times so that $n_e \simeq 2.2 \times 10^{-4} n_H$ where the hydrogen number density is obtained from the virial density of (2.53) as follows:

$$n_H = \frac{\varrho_{vir}}{\mu m_p} \frac{\Omega_b}{\Omega_M} \tag{3.9}$$

where $\mu \simeq 1.087(\bar{m}/m_p) \simeq 1.351$ is the mean gas mass per hydrogen atom, m_p the proton mass and $\bar{m}$ is the mean molecular mass. The coefficient 1.351 accounts for the presence of helium contributing to the gas mass and for the fractional number density of hydrogen. Replacing the expression of ϱ_{vir} from (2.53) we find:

$$n_H = 0.97 \left(\frac{1+z}{31} \right)^3 \tag{3.10}$$

Making use of (2.14) for the residual ionized fraction x and (3.10) we find that $n_e = x n_H$ is:

$$n_e = 2.1 \times 10^{-4} \left(\frac{1+z}{31}\right)^3 \qquad (3.11)$$

Finally, substituting in (3.7) f_{H_2} with the expression for the minimum H_2 fraction needed from collapse from (2.62), and making use of (3.11) and (3.8), we find the H_2 formation timescale:

$$\tau_{form} = 3.23 \times 10^{13} \text{ s} \left(\frac{M}{10^6 M_\odot}\right)^{-2.2187} \left(\frac{1+z}{31}\right)^{-7.828} \qquad (3.12)$$

The relation between LW background and minimum halos mass can be found by equating the two timescales from (3.6) and (3.12). Thus, we obtain:

$$J_{21} = 0.028 \left(\frac{M}{10^6 M_\odot}\right)^{2.2187} \left(\frac{1+z}{31}\right)^{7.828} \qquad (3.13)$$

In order to derive this expression we have ignored the significant departure from homogeneity in the halo. The effect of the significant variations in the gas physical properties as a function of radius and the additional possible presence of shocks will have a major effect in the capacity of a LW background to suppress collapse [193, 297]. Often the suppression will turn into a delay of the collapse even though clearly a long delay is for all practical purposes a suppression as the original halo will gain significant mass during such delay. We can compare the results from our simple analysis with those of more elaborate calculations. Expressed in terms of the critical mass as a function of J_{21}, (3.13) becomes:

$$M_c = 5.04 \times 10^6 M_\odot J_{21}^{0.451} \left(\frac{1+z}{31}\right)^{-3.528} \qquad (3.14)$$

For $J_{21} = 0$ the expression from (3.14) goes to zero but the minimum mass is set to $1.36 \times 10^5 M_\odot$ by (2.63). This could be compared with the expression from Machacek et al. [148] derived by parametrizing the results of their numerical calculations and expressed in terms of J_{21} and computed for a 4% fraction of gas available for star formation [297]:

$$M_M = 2.5 \times 10^5 + 5.59 \times 10^6 J_{21}^{0.47} M_\odot \qquad (3.15)$$

Thus, we see that the dependence from J_{21} is very similar in both cases except for the explicit inclusion in Machacek et al.'s expression of the critical mass for $J_{21} = 0$ and for the redshift dependence of (3.14). Detailed numerical simulations [123, 297] are in excellent agreement with (3.15) but are also in very good agreement with (3.14) for the case when J_{21} is sufficiently large that the critical mass for $J_{21} = 0$ can be neglected.

We have seen that the background to effectively suppress Population III formation in $10^6 M_\odot$ halos is larger than that required to photodissociate primordial H_2,

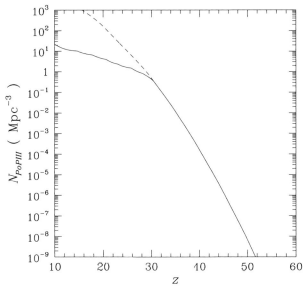

Fig. 3.4 Cumulative number density of Population III stars per comoving Mpc^3 as a function of redshift considering the effect of the Lyman–Werner background buildup. The solid line is computed including the effect of the LW background, while the dashed line corresponds to the nonbackground case and is shown here for reference. Both have been derived considering halos with a minimum mass of $6 \times 10^5 M_\odot$.

e.g. $J_{21} \simeq 0.03$ for a $10^6 M_\odot$ halo compared to $\sim 1.6 \times 10^{-3}$ to dissociate all primordial H_2. Thus, while a single Population III star emits sufficient LW photons to effectively pre-empt H_2 cooling in its vicinity, the build up of a sufficient background to suppress all halos will require more time. In practice, halos are not static objects and a halo will gain mass with time so that what is effectively suppressed at some redshift will not stay suppressed indefinitely. However, we can capture this effect by considering halos of larger mass. We can compute a new cumulative density of Population III stars by keeping track of the background build up and suppressing star formation in small halos as necessary. The result is shown in Figure 3.4 where we plot the cumulative number density of Population III stars formed in halos more massive than $6 \times 10^5 M_\odot$ and for comparison the same quantity computed ignoring the LW background buildup. The smallest halos are the most vulnerable to LW feedback and clearly at redshift $z \simeq 30$ they begin to be suppressed and Population III star formation continues only in the more massive ones. It is worth stressing that this estimate is rather crude and suffers in particular from two major approximations: one is that more massive halos may well have been polluted by metals generated by previous generations of stars so that they would indeed form stars but they would not be Population III objects. The second approximation is that we have simply counted LW photons and not carried out a radiative-transfer calculation. This in practice leads to ignoring the detailed redshifting of photons in and out of the LW bands that we treat only statistically.

3.2.3
Late Population III Star Formation by Atomic-Hydrogen Cooling in Massive Halos

The standard picture is that once the Lyman–Werner background is established and further formation of Population III stars by molecular-hydrogen cooling is suppressed, Population III stars can be formed only through atomic-hydrogen cooling in halos with mass of $10^8 M_\odot$ [123]. A first difficulty of this simple scheme is that halos with mass $>10^6 M_\odot$ are very effective at self-shielding against the LW background (see Section 3.2.2). Moreover, in order to preserve pristine material in a $10^8 M_\odot$ halo it is necessary that star formation in the smaller halos contained in the more massive one be suppressed by the LW background. In many cases this will not be the case and the more massive halos will have been polluted by a Population III star and will no longer be able to form another object of primordial metallicity. Figure 3.5 shows that a $10^8 M_\odot$ solar mass halo collapsing at redshifts of 10 to 20 was most likely enriched by previous generations of Population III stars forming in halos of mass $10^6 M_\odot$ collapsing at higher redshifts.

Indeed, simulations of a halo of $5 \times 10^7 M_\odot$ collapsing at $z = 10$ show that about 10 Population III.1 and several Population III.2 form in subhalos before the halo can cool by atomic hydrogen [102]. Whether or not this happens and pristine stars

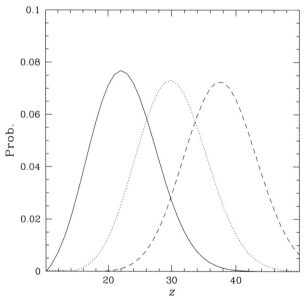

Fig. 3.5 Probability distribution that a halo of $10^8 M_\odot$ at $z = 10$ (solid line), 14 (dotted line), 18 (dashed line) had a mass of $10^6 M_\odot$ at the given redshift. It is clear that for halos at $z = 14$ and 18 it is very likely that they had a mass of $10^6 M_\odot$ at a redshift sufficiently high that a Lyman–Werner background was not in place and they could form Population III stars capable of polluting the larger halos including them at later redshift. For $10^8 M_\odot$ halos at $z = 10$ there is a significant probability that they might not be polluted because their progenitor on the mass scale $10^6 M_\odot$ could be prevented from forming Population III stars by an established Lyman–Werner background.

are formed at $z = 10$ depends on our assumption about cooling and mixing of chemically enriched material. Similarly halos of mass $3.5 \times 10^7 M_\odot$ collapsing at $z \simeq 15$ host 10–26 Population III stars early in their assembly and are enriched to $\sim 10^{-3} Z_\odot$ [298].

3.2.4
Termination of the First Stars Phase

It is clear from the discussion so far that while radiative feedback and the establishment of a LW background are capable of moderating and self-regulating Population III star formation, they cannot completely suppress it and formation of stars with primordial chemical composition can continue in more massive halos as long as pristine material is available. Thus, the termination of the first stars phase must be due to chemical pollution rather than radiative feedback.

3.3
Containing Gas in the Halos of Population III Stars

In this section we will study the conditions under which gas heating by ionization or by supernova explosions can be retained within a halo and we will also discuss the physical consequences on the evolution of subsequent generations of halos.

3.3.1
Ionization Heating and Gas Temperature

A Population III star is able to ionize all diffuse gas in the host halo. Dense knots capable of forming another Population III star may be self-shielded due to their enhanced recombination rate but small halos tend to form a single star (see also Section 3.4) and we can thus ignore the presence of extra cores here.

Ionizing sources have a spectral energy distribution extending smoothly beyond the Lyman limit and, as a consequence, they will have a mean energy of their ionizing photons exceeding by some finite amount the hydrogen ionization energy. For a star with spectral energy distribution $L(\nu)$ we can compute the excess energy per ionization as:

$$\varepsilon_{ex} = \frac{\int_{\nu_c}^\infty h(\nu - \nu_c) \frac{L(\nu)}{h\nu} d\nu}{\int_{\nu_c}^\infty \frac{L(\nu)}{h\nu} d\nu} \tag{3.16}$$

The expression above is valid as long as the column density of neutral hydrogen is high enough that most photons lead to ionization regardless of their energy as long as it is above $h\nu_c$. For lower column density a more relevant quantity is the excess energy weighted by the cross section for hydrogen ionization σ_I, namely:

$$\varepsilon_{wgt} = \frac{\int_{\nu_c}^\infty h(\nu - \nu_c) \frac{L(\nu)}{h\nu} \sigma_I(\nu) d\nu}{\int_{\nu_c}^\infty \frac{L(\nu)}{h\nu} \sigma_I(\nu) d\nu} \tag{3.17}$$

Tab. 3.1 Mean excess energy ε_{ex} and mean weighted excess energy ε_{wgt} of ionizing photons for various blackbody temperatures.

B.B. Temperature (K)	ε_{ex} (eV)	ε_{wgt} (eV)
3×10^4	3.51	2.33
5×10^4	6.70	3.66
10^5	15.95	6.33
1.1×10^5	18.03	6.78

The ionization cross section is lower at higher energy so we expect the weighted excess energy given by (3.17) to be lower than that given by the unweighted expression of (3.16). For the ionization cross section we use the expression given by Osterbrock [196]:

$$\sigma_I(\nu) = S_0 \frac{\exp\left(4 - \frac{4\tan^{-1}\sqrt{\nu/\nu_c - 1}}{\sqrt{\nu/\nu_c - 1}}\right)}{1 - \exp\left(-\frac{2\pi}{\sqrt{\nu/\nu_c - 1}}\right)} \left(\frac{\nu_c}{\nu}\right)^4 \tag{3.18}$$

where the constant $S_0 = 6.3 \times 10^{-18}$ cm².

In practice, for a realistic situation, the effective excess energy will be intermediate between the two values. Initially the low-energy photons begin ionizing but as they do so the effective spectrum becomes harder with the high-energy photons being progressively responsible for more ionizations. We can compute the two values of excess energy for a series of blackbody spectra at various temperatures. The values are given in Table 3.1.

The excess energy is transferred to the electrons. If the system was far from equilibrium and one could ignore recombinations, the excess energy would directly give us the electron temperature or, in the case of equipartition, the gas temperature. However, in the denser halo gas recombinations cannot be neglected and one has to compute the gas temperature in the presence of an equilibrium between recombinations and ionizations.

Let us compute the gas equilibrium temperature for a pure hydrogen gas. Following Osterbrock [196] the heating term can be written in terms of the rate of generation of excess energy due to ionizations, namely:

$$n_{HI} \int_{\nu_c}^{\infty} h(\nu - \nu_c) \frac{J_\nu}{h\nu} d\nu \tag{3.19}$$

where n_{HI} is the neutral hydrogen density and J_ν is the volume emissivity at the location under consideration. The coefficient α_A is the total recombination coefficient to all levels including the ground state. It can be obtained by summing over all levels, adding recombinations to the ground level to α_B or from interpolation [1] and in the simplest form it is given – to a very good approximation for the temper-

3.3 Containing Gas in the Halos of Population III Stars

atures of interest – by:

$$\alpha_A \simeq 4.17 \times 10^{-13} \left(\frac{T}{10^4 \text{ K}}\right)^{-0.724} \tag{3.20}$$

Note that we use the case A value of the recombination coefficient, α_A, rather than α_B because the gas is heated by each ionization regardless of whether it is followed by a recombination to an excited state or to the ground state. The radiative equilibrium condition is given by:

$$n_{HI} \int_{\nu_c}^{\infty} \frac{4\pi J_\nu}{h\nu} \sigma_I(\nu) \, d\nu = n_e n_p \alpha_A \tag{3.21}$$

where the integral in the left-hand side is the number of ionizations for a volume emissivity J_ν. Making use of (3.21) to derive n_{HI} and substituting it in (3.19) we find the expression for the heating term:

$$\dot{E}_{heat} = n_e n_p \alpha_A \frac{\int_{\nu_c}^{\infty} h(\nu - \nu_c) \frac{J_\nu}{h\nu} \sigma_I(\nu) \, d\nu}{\int_{\nu_c}^{\infty} \frac{J_\nu}{h\nu} \sigma_I(\nu) \, d\nu} \tag{3.22}$$

In the case where the volume emissivity has the same spectral shape as the star spectral energy distribution, e.g., if $J_\nu = L(\nu)/4\pi R^2$, the integral in the expression above becomes the weighted excess energy of (3.17) and (3.22) becomes:

$$\dot{E}_{heat} = n_e n_p \alpha_A \varepsilon_{wgt} \tag{3.23}$$

In realistic situations, the radiation spectrum will be modified by ionization and J_ν will cease to be proportional to $L(\nu)$. In the limiting case where all ionizing photons are absorbed and when applied to the average conditions in the cloud one will replace ε_{wgt} in (3.23) with ε_{ex} given by (3.16).

In order to properly evaluate cooling we need to consider four contributions:

- **Recombination cooling.** When an electron recombines, the energy corresponding to the relevant energy level is radiated away. Recombination cooling is given by [9]:

$$\dot{E}_{rec} \simeq 5.49 \times 10^{-25} \text{ erg cm}^3 \text{ s}^{-1} \left(\frac{T}{10^4 \text{ K}}\right)^{0.3} \left[1 + 0.04 \left(\frac{T}{10^4 \text{ K}}\right)^{0.7}\right]^{-1} n_e n_p \tag{3.24}$$

- **Bremsstrahlung.** This is also known as free–free cooling [196] and is given by [9]:

$$\dot{E}_{brem} \simeq 1.43 \times 10^{-27} \text{ erg cm}^3 \text{ s}^{-1} \sqrt{T} \left[1.1 + 0.34 e^{-(5.5 - \log_{10} T)^{2/3}}\right] n_e n_p \tag{3.25}$$

- **Collisional excitation cooling.** This is also known as Lyman α cooling. When a neutral atom is collisionally excited, the energy transferred in the collision can be radiated away by emission of, e.g., Lyman α photon or photons associated to other transitions. The rate is given by [9]:

$$\dot{E}_{cec} \simeq 7.5 \times 10^{-19} \text{ erg cm}^3 \text{ s}^{-1} \left[1 + \sqrt{\left(\frac{T}{10^5 \text{ K}}\right)}\right]^{-1}$$

$$e^{-\frac{118348}{T}} n_e^2 n_p \frac{\alpha_A}{\int_{\nu_c}^{\infty} \frac{4\pi J_\nu}{h\nu} \sigma_I(\nu) d\nu}$$
(3.26)

where we have replaced n_{HI} with its expression derived from (3.21).

- **Collisional ionization cooling.** This is analogous to the collisional excitation cooling but in this case the collision leads to an ionization. Clearly this process depends on the collisional ionization rate [1] k_1 and also depends on the density of neutral hydrogen. It can be expressed as follows [9]:

$$\dot{E}_{cic} \simeq 2.18 \times 10^{-11} \text{ erg cm}^3 \text{ s}^{-1} n_e^2 n_p k_1 \frac{\alpha_A}{\int_{\nu_c}^{\infty} \frac{4\pi J_\nu}{h\nu} \sigma_I(\nu) d\nu}$$
(3.27)

where for k_1 one can use the interpolation given by Abel et al. [1]. Below, we give for convenience this interpolation truncated to the first 5 terms, which is sufficient for the temperatures that we are interested in:

$$\ln k_1 \simeq -32.71396786 + 13.536556 \ln T - 5.73932875 (\ln T)^2 \\ + 1.56315498 (\ln T)^3 - 0.2877056 (\ln(T))^4$$
(3.28)

The gas equilibrium temperature can be found by requiring that $\dot{E}_{heat} = \dot{E}_{rec} + \dot{E}_{brem} + \dot{E}_{cec} + \dot{E}_{cic}$. In order to solve this equation we need to assume a value for the density. In fact most of the terms depend on the product $n_e n_p$ but collisional excitation cooling and collisional ionization cooling depend on $n_e^2 n_p$. It is easy to see that for the conditions of interest, e.g., at the half-mass radius in a $10^6 M_\odot$ halo containing a $300 M_\odot$ star, the fraction of neutral hydrogen is always small ($\approx 10^{-4}$) and thus we can assume for the electron density n_e the mean number density given by (3.10). The results for a number of blackbody temperatures are given in Table 3.2. In addition to the value of temperature T_{gas} derived for the excess energy obtained by weighting with the ionization cross section we give also the unweighted value T_{ex} to illustrate how at high temperatures, a change of a factor 3 in excess energy causes only a change by 30% in equilibrium temperature. We have verified that these results for T_{gas} are in good agreement with the volume-averaged electron temperature in a HII regions with primordial metallicity – and therefore including helium – as calculated by Cloudy [82].

In the calculation of the equilibrium temperature we have ignored an additional cooling mechanism: inverse compton cooling over the cosmic microwave background radiation (see (4.26)). This effect strongly depends on redshift and it is easy to verify [9] that in the conditions of interest and at $z \lesssim 30$ this mechanism only provides a small ($\lesssim 5\%$) reduction of the temperatures given in Table 3.2. The reduction grows up to $\lesssim 30\%$ at $z = 50$.

Tab. 3.2 Equilibrium temperature computed for the mean excess energy T_{ex} and for the mean weighted excess energy T_{gas}.

B.B. Temperature (K)	T_{ex} (K)	T_{gas} (K)
3×10^4	22 500	18 250
5×10^4	27 500	22 000
10^5	32 500	24 250
1.1×10^5	33 000	24 500

3.3.2
The Escape of Gas Heated by Ionization

We now derive the minimum mass of a halo able to retain gas heated by ionization. First, let us determine the maximum mass of a halo that is fully ionized. The number of hydrogen atoms in a halo with total mass M is given by:

$$N_H = \frac{M}{\mu m_p} \frac{\Omega_b}{\Omega_m} \simeq 1.66 \times 10^{62} \left(\frac{M}{10^6 M_\odot} \right) \tag{3.29}$$

where we account for the presence of helium by correcting the proton mass by $\mu = 1.35$. The number of ionizing photons is:

$$N_\gamma = Q_0 M_\star \tau_\star \simeq 3.7 \times 10^{64} \left(\frac{M_\star}{300 M_\odot} \right) \tag{3.30}$$

where we have adopted $Q_0 = 1.3 \times 10^{48}$ s^{-1} M$_\odot^{-1}$ from Table 2.2 and the lifetime from (2.79).

By equating $N_H = N_\gamma$ we find that the maximum mass of a halo fully ionized by a $300 M_\odot$ Population III star is $M \simeq 2.22 \times 10^8 M_\odot$. This number has been derived while neglecting recombinations and thus it is likely that the true maximum mass will be lower.

In order to consider the effect of recombinations we can derive the radius of the Stroemgren sphere, namely the sphere inside which recombinations and ionizations are balanced. For a homogeneous system we can write the balance of recombinations and ionizations as:

$$\frac{4\pi}{3} R_s^3 n_e n_p \alpha_B = Q_0 M_\star \tag{3.31}$$

Here, we use the case B recombination coefficient because recombinations to the ground state lead to another ionization and this balances out in this equation. In (3.31), R_s is the Stroemgren radius that we can write as:

$$R_s = \left(\frac{3}{4\pi} \frac{Q_0 M_\star}{n^2 \alpha_B} \right)^{1/3} \simeq 75 \text{ pc} \left(\frac{1+z}{31} \right)^{-2} \tag{3.32}$$

where we have used $\alpha_B = 1.63 \times 10^{-13}$ – applicable to $T = 2 \times 10^4$ K – and the fact that the gas is highly ionized within the Stroemgren sphere to replace n_e and n_p

with n. The correct value of n is obtained from the density prior to collapse given in (3.10). Following the discussion in Section 2.3.3, the halo temperature decreases by a factor $\simeq 6.97$ in order to reach the Jeans instability. In the isobaric approximation the density will increase by a similar factor to yield the mean number density after collapse:

$$n_H \simeq 6.76 \text{ cm}^{-3} \left(\frac{1+z}{31}\right)^3 \tag{3.33}$$

The total ionized gas mass is obtained by multiplying the volume of the Stroemgren sphere by the gas density. The corresponding halo mass is obtained by multiplying the ionized gas mass by Ω_m/Ω_b, so that:

$$M_s \simeq 2.31 \times 10^6 M_\odot \left(\frac{1+z}{31}\right)^{-3} \left(\frac{M_\star}{300 M_\odot}\right) \tag{3.34}$$

As expected, this value is much smaller than that obtained when ignoring recombinations. From (3.34) we can see that Population III stars with mass $M_\star \lesssim 130 M_\odot$ are not able to fully ionize a halo of $10^6 M_\odot$. In Section 4.3.1 we will describe the implications of these results on the escape fraction of ionizing photons for Population III stars.

Let us now determine the minimum halo mass able to contain the ionized gas. If this mass is lower than the mass that can be completely ionized the implication is that halos more massive than this limit will be completely ionized but the gas will remain bound. In the opposite case we would have to repeat the calculation for partially ionized halos.

Let us focus on a fully ionized halo and assume a temperature of 2×10^4 K for the gas. This temperature corresponds to a thermal velocity of the protons:

$$v_{th} = \sqrt{\frac{3kT}{m_p}} \simeq 22.25 \text{ km s}^{-1} \left(\frac{T}{2 \times 10^4 \text{ K}}\right)^{1/2} \tag{3.35}$$

In order to estimate when this gas can be contained within a halo we need to determine the escape velocity. This is given by:

$$v_{esc} = \sqrt{\frac{2GM}{R_h}} \tag{3.36}$$

where the half-mass radius R_h is given by (2.54) and (2.55) and equals:

$$R_h \simeq 86 \text{ pc} \left(\frac{M}{10^6 M_\odot}\right)^{1/3} \left(\frac{1+z}{31}\right)^{-1} \tag{3.37}$$

Replacing (3.37) into (3.36), we find the escape velocity:

$$v_{esc} \simeq 10 \text{ km s}^{-1} \left(\frac{M}{10^6 M_\odot}\right)^{1/3} \left(\frac{1+z}{31}\right)^{1/2} \tag{3.38}$$

3.3 Containing Gas in the Halos of Population III Stars

Tab. 3.3 Minimum halo mass able to confine the hot gas ionized by a single $300 M_\odot$ Population III star as a function of redshift.

Redshift (z)	M_{conf} ($M_\odot$)
10	9.9×10^7
20	6.4×10^6
30	4.8×10^6
40	3.9×10^6
50	3.3×10^6

Thus, gas will be able to escape when the thermal velocity of (3.35) is higher than the escape velocity. This translates into a condition on the halo mass. The gas can escape when the halo mass is lower than M_{esc}, which is given by:

$$M_{esc} = 1.1 \times 10^7 M_\odot \left(\frac{1+z}{31}\right)^{-3/2} \tag{3.39}$$

This mass is larger than that of the Stroemgren sphere given by (3.34). This implies that for masses lower than that of the Stroemgren sphere, halos become fully ionized and are incapable of confining the hot gas. For masses above M_{esc} the gas will be confined. For intermediate masses the hot gas could also be confined as the expanding ionized gas mixes with neutral gas in the halo. Dilution with neutral gas will reduce the mean thermal velocity of the gas. The average thermal velocity squared in a halo with mass $M > M_s$ is given by:

$$\langle v^2 \rangle = \frac{\frac{\Omega_b}{\Omega_m}\left[M_s v_{th}^2 + (M - M_s)v_{vir}^2\right]}{\frac{\Omega_b}{\Omega_m} M}$$

$$\simeq 22.25^2 \frac{M_s}{M} + 7.09^2 \left(\frac{M}{10^6 M_\odot}\right)^{2/3} \left(\frac{1+z}{31}\right)\left(1 - \frac{M_s}{M}\right) \text{ km}^2 \text{ s}^{-2} \tag{3.40}$$

where we have used $v_{vir} = \sqrt{3kT_{vir}/\bar{m}}$ with T_{vir} given by (2.57). The quantities Ω_b/Ω_m are used to convert the halo mass to baryon mass, however, they cancel out, given that we keep constant the ratio of baryons to total mass. This is not strictly valid as the baryon fraction of halos is seen to vary in simulations because of feedback [298]. Requiring that $\sqrt{\langle v^2 \rangle} = v_{esc}$ now gives us a confinement mass $M_{conf} \simeq 4.8 \times 10^6 M_\odot$ for $z = 30$. The values calculated for other redshifts are given in Table 3.3.

The calculations leading to the minimum mass M_{conf} are based on pure energy arguments. The actual evolution of a halo will be more complex. An ionization front (I-front) will first form. The front velocity can be derived from the ionizing flux and the number density of the halo. At the half-mass radius the ionizing flux per unit area is given by $Q_0 M_\star/4\pi R_h^2$. The mean density of the halo after collapse is given by (3.33). The I-front propagation velocity is thus given by:

$$v_I = \frac{Q_0 M_\star}{4\pi R_h^2 n_H} \simeq 597 \text{ km s}^{-1} \left(\frac{M}{10^6 M_\odot}\right)^{-2/3} \left(\frac{1+z}{31}\right)^{-1} \qquad (3.41)$$

Note that in a realistic system the density is not uniform and in fact the radial distribution found in simulations is close to that of an isotherm sphere with $\varrho \propto r^{-2}$. This implies that (3.41) remains largely valid as both the ionizing photon flux and the density will vary as r^{-2}. The I-front velocity has to be compared to the isothermal sound speed:

$$c_{iso} = \frac{2kT}{m_p} \simeq 18.2 \text{ km s}^{-1} \left(\frac{T}{2 \times 10^4 \text{ K}}\right)^{1/2} \qquad (3.42)$$

Since the I-front velocity is much larger than the sound speed it is to be expected that the I-front will leave behind a mechanically unbalanced system that will not be able to adjust fast enough to the new temperature distribution behind the I-front. The result is going to be the creation of a shock front propagating outwards. Detailed calculations show that this shock front propagates with velocities of the order of 30 km s^{-1} [291]. The values of the minimum mass M_{conf} that we have derived are in very good agreement with the results of simulations [291].

3.3.3
The Escape of Gas Following a Supernova Explosion

We have seen in Section 2.3.5 that Population III stars with mass between 140 and 260$M_\odot$ can end their life as a pair-instability supernova (PISN). The brightest of these supernovae, at the high mass end of the range, has a kinetic energy output of 9.35×10^{52} erg [113]. This energy, when distributed over the baryonic mass of a halo of total mass M, gives a thermal velocity:

$$v_{th} = 232 \text{ km s}^{-1} \left(\frac{M}{10^6 M_\odot}\right)^{-1/2} \qquad (3.43)$$

As we did previously for the ionization heating case, we can compare this thermal velocity to the escape velocity of the halo given by (3.38) to find the minimum halo mass able to contain the supernova heated gas. Comparing the two velocities gives:

$$M_{PISN} \simeq 4.35 \times 10^7 M_\odot \left(\frac{1+z}{31}\right)^{-3/5} \qquad (3.44)$$

In practice, the real value of M_{PISN} will be lower than the value given in (3.44) because some of the SN energy energy will be radiated away. Halos with mass exceeding M_{PISN} will be able to contain the gas heated by the supernova and will be enriched in metals. Lower-mass halos will disperse the metals into the IGM. In a PISN, most of the stellar mass is converted to elements heavier than helium, i.e. 'metals' in the ordinary jargon of astronomers. In the specific case about $127 M_\odot$ of metals are released [113], enriching a halo with mass M_{PISN} to $1.5 \times 10^{-4} Z_\odot$. The estimate of (3.44) ignores cooling. We should expect that when cooling is included the minimum mass will be lower than the value we give in (3.44) and consequently the minimum mass halo will be enriched to higher metallicity. Clearly, as in the case of the ionization-heated gas, the correct calculation would require modeling the propagation of a shock through the halo gas (see Figure 3.6). However, even in this case the results would be comparable.

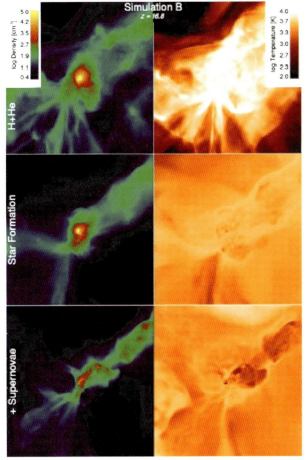

Fig. 3.6 Illustration of the effect of feedback due to star formation and supernova explosions in a primordial halo at $z = 16.8$ [298]. The left panels show and the gas density and the right panels the temperature. The top panels refer to the case when only hydrogen and helium cooling are present. In the intermediate panels star formation is also included and leads to more substructure and – unintuitively – lower temperatures. The bottom panel illustrates the effect of supernova explosions (Reproduced by permission of the AAS).

3.3.4
Population II.5

Halos with mass exceeding M_{PISN} given in (3.44) will be able to retain gas and will be enriched in metals. Metallicity generally makes cooling more effective. For the more-massive halos with virial temperatures above 10^4 K, metallicity needs to exceed $10^{-3} Z_\odot$ in order to affect cooling (see, e.g., Figure 3.7). Stars forming from material with metallicity below $\approx 10^{-3} Z_\odot$ will still form similarly to Population III stars [120, 245] and will have very high effective temperatures. However, we expect that these objects may have a more normal initial mass function as fragmentation is more likely for nonzero metallicity. Moreover, at lower redshift more substructure is present around a halo and this increases the chances that torques will generate angular momentum, which in turn is also known to favor fragmentation [58]. Thus, it is likely that Population II.5 objects will not be single stars but small clusters of lower-mass objects, opening up the possibility of finding them in searches of very low mass stars in the Galaxy.

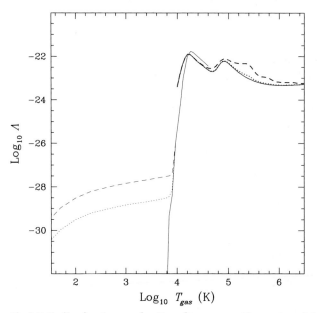

Fig. 3.7 Cooling function as a function of temperature and chemical composition. The thick lines are from Sutherland and Dopita [269], while the thin lines have been derived following Bromm *et al.* [40]. For $T \gtrsim 10^4$ K the two sets of curves are only in approximate agreement. We do not include molecular hydrogen and consider a primordial composition (solid line), a metallicity of $10^{-3} Z_\odot$ (dotted line) or a metallicity of $10^{-2} Z_\odot$ (dashed line). By comparison to Figure 2.4, it is clear that for metallicity $10^{-3} Z_\odot$ or lower cooling below 10^4 K is going to be dominated by molecular hydrogen as for the primordial metallicity case.

3.4
The First Star Clusters

In Section 2.3.3 we derived the conditions for collapse and assumed that one star per halo would be formed. Here, we want to revisit the physical conditions at collapse to determine whether multiple stars can actually be formed in a sufficiently massive halo. A necessary – but not sufficient – condition is that the cooling time, at the moment when a halo has reached the Jeans instability condition, be shorter than the dynamical collapse time. If this condition is not satisfied, the Jeans collapse will proceed over the whole system. In contrast, when the condition is satisfied, there is the possibility for continued efficient cooling over timescales shorter than the Jeans collapse. Such a process could reduce the Jeans mass to a fraction of the halo baryonic mass, possibly leading to the formation of several stars. We start by determining the condition for the cooling time to be shorter than the collapse time at the moment when the halo is about to become Jeans unstable.

The cooling time τ_{cool} is given in terms of T and $\Lambda(T)$ by (2.58). Replacing the halo conditions at collapse, one obtains an expression for τ_{cool} as a function of halo mass and redshift (2.59). However, we have seen in Section 2.3.3 that the temperature at the moment of the global Jeans instability has decreased by a factor 6.97 with respect to the collapse value. In the isobaric approximation the product of density and temperature is constant so that the density will have increased by a factor 6.97. Thus, at the moment of the Jeans instability we have:

$$T_J = T_{vir}/6.97 = 362 \text{ K} \left(\frac{M}{10^6 M_\odot}\right)^{2/3} \left(\frac{1+z}{31}\right) \tag{3.45}$$

and the number density n_H is given by (3.33).

Replacing these values into the expression of (2.58) taking into account the approximation for $\Lambda(T)$ of (2.32), we find:

$$\tau_{cool,J} = 5.5 \times 10^{11} \text{ s} \left(\frac{M}{10^6 M_\odot}\right)^{-1.6} \left(\frac{1+z}{31}\right)^{-5.4} f_{H_2}^{-1} \tag{3.46}$$

Analyzing the dependence of τ_{cool} to temperature and in the isobaric approximation we find that $\tau_{cool} \propto T^{1.4}$ and indeed the difference between (3.46) and (2.59) giving the cooling time before the halo cools is equal to $6.97^{1.4}$.

We now need to derive an expression for the collapse timescale, which we take to be $\tau_{coll,J} = 1/\sqrt{4\pi G \varrho_b}$. Using for ϱ_b the expression $6.97 \Omega_b \varrho_{vir}/\Omega_m$, and (2.53) for ϱ_{vir}, we find:

$$\tau_{coll,J} = 2.783 \times 10^{14} \text{ s} \left(\frac{1+z}{31}\right)^{-3/2} \tag{3.47}$$

Requiring the timescales in (3.46) and (3.47) to be identical gives us a condition on the required H_2 fraction:

$$f_{H_2} = 1.98 \times 10^{-3} \left(\frac{M}{10^6 M_\odot}\right)^{-1.6} \left(\frac{1+z}{31}\right)^{-3.9} \tag{3.48}$$

We can turn (3.48) into a condition on the mass by using (2.61) for the maximum (asymptotic) H_2 fraction as a function of mass. This gives:

$$M_{fragm} \simeq 10^6 M_\odot \left(\frac{1+z}{31}\right)^{-2.07} \tag{3.49}$$

For halos with $M > M_{fragm}$, the conditions at the Jeans instability are such that the cooling time is shorter than the gravitational collapse time. Thus, these halos could continue to cool and possibly form multiple stars. For halos with $M < M_{fragm}$ the cooling time is longer than the collapse time and it is to be expected that these halos will form a single Population III star. This is in broad agreement with the results of numerical simulations [38].

The condition for fragmentation given by (3.49) is necessary but not sufficient. One additional requirement on a halo is that the minimum Jeans mass achieved before H_2 cooling becomes too inefficient has to be a few times smaller than the total baryonic mass. This 'multiplicity' condition would at least guarantee that multiple Jeans instabilities could take place simultaneously in the halo, possibly leading to the formation of a star cluster. We can derive the multiplicity condition similarly to the derivation leading to (2.68). Indeed, this equation was derived by requiring that the Jeans mass was identical to the total baryonic mass. A condition for a multiplicity N_s is obtained by requiring that the Jeans mass M_J of (2.66) satisfies the condition;

$$M_J N_s = 0.173 M \tag{3.50}$$

or, by using the temperature change condition given by (2.67):

$$M_{mult} = 1.905 \times 10^5 N_s^{3/2} M_\odot \left(\frac{1+z}{31}\right)^{-3/2} \tag{3.51}$$

Equation (3.51) can be cast in terms of the maximum multiplicity that a halo satisfying the fragmentation mass condition of (3.49) can achieve:

$$N_s \lesssim 3.02 \left(\frac{M}{10^6 M_\odot}\right)^{2/3} \left(\frac{1+z}{31}\right) \tag{3.52}$$

Equation (3.52) limits the maximum number of stars that a halo able to fragment will form. For a halo with $M = 10^8 M_\odot$ collapsing at $z = 20$ we find $N_s \lesssim 44$.

The accuracy of (3.52) can be tested with numerical simulations. Bromm and collaborators [38] studied the collapse of a number of halos with mass $2 \times 10^6 M_\odot$ at $z = 30$ and found that in general several clumps were forming. For the low and intermediate angular momentum cases the number of clumps was 2 to 5, which compares well with $N_s \simeq 4.79$ obtained from (3.52). In the case of high angular momentum several lower-mass clumps would form. Our analysis completely ignores the effects of angular momentum so it is not surprising that when angular momentum is important it leads to different results. Halos with mass of $2 \times 10^5 M_\odot$ form instead a single clump, as expected since this mass is below the fragmentation mass of (3.49).

Interestingly, although less directly related, is the comparison with the results by Greif et al. [102] who studied the assembly history of a halo with mass $5 \times 10^7 M_\odot$ at $z = 10$ and found that halos merging into it form 10 Population III.1 stars and a number of Population III.2. This number of Population III.1 stars is compatible with the value $N_s \simeq 14.54$ from (3.52) but it should be stressed that in the simulations these Population III.1 stars formed before the halos had merged together into the $5 \times 10^7 M_\odot$ halo so that the outcome is in this case more dictated by dark halo merging statistics then by Jeans instability properties.

The multiplicity condition in (3.52) was derived by making use of (2.67) that assumes a minimum temperature for molecular-hydrogen cooling of 120 K. We can estimate the minimum metallicity for which the cooling function of a nonprimordial system exceeds the molecular-hydrogen cooling function at $T = 120$ K for a molecular-hydrogen fraction of $f_{H_2} = 10^{-4}$. Using the approximation $\mathrm{Log}_{10}\Lambda(120\ \mathrm{K}) \simeq -29.3 + \mathrm{Log}_{10}(Z/10^{-3} Z_\odot)$ [40], we find $Z \simeq 5 \times 10^{-6}$. Thus, even at very low metallicity additional fragmentation and the formation of smaller mass stars is, in principle, possible.

3.4.1
Clusters of Population III Stars and of Metal-Poor Stars

As we have seen in Section 3.2.3 the probability of a $10^8 M_\odot$ halo collapsing before being chemically polluted is very low so that the probability of forming a cluster of a few tens of Population III stars is relatively low. The main factor at play is that it is very unlikely that a perturbation of $10^8 M_\odot$ collapses without some perturbation of lower mass having collapsed and formed Population III stars previously. The most likely scenario for this to happen is if the collapse of the $10^8 M_\odot$ perturbation occurs relatively late when a Lyman–Werner background is in place. This ensures that the lower-mass perturbations that would have otherwise formed a previous generation of Population III stars are inhibited by radiative feedback. This is illustrated in Figure 3.5. Following the analysis in Section 3.4 we would expect such a primordial star cluster forming at redshift $z \simeq 10$ to contain a few tens of very massive stars.

However, it is likely that these star clusters of primordial stars will not be the highest redshift star clusters as the majority of the objects forming following the collapse of a $10^8 M_\odot$ halo have been enriched to low metallicity [298], e.g., $z \simeq 10^{-3} Z_\odot$ and will form low-metallicity star clusters. These clusters will be made of lower-mass stars but may still be detectable to high redshifts.

Objects forming stars in a halo with a total mass of $10^8 M_\odot$ are sometimes referred to as dwarf galaxies in the literature. Here, we refer to these objects as star clusters because their stellar mass is comparable to that of star clusters in the Milky Way. We can estimate their size as follows. The half-mass radius from (2.54) and (2.55) can be written as:

$$R_h = 1125\ \mathrm{pc} \left(\frac{M}{10^8 M_\odot}\right)^{1/3} \left(\frac{11}{1+z}\right) \tag{3.53}$$

The radius of the collapsed system will be at least a factor 2 smaller (e.g. cf. with the factor 6.97 for the increase in density discussed in Section 3.4) so the object radius will be ~500 pc, intermediate between present-day galaxies and star clusters.

3.4.2
The Origin of Globular Clusters

Peebles and Dicke [208] speculated in 1968 that globular clusters were the first objects to collapse in the Universe. While the specific cosmological parameters considered were very different from those generally accepted today, the formation of globular clusters remains a very interesting problem. Indeed, despite evidence that some globular clusters form in the Universe today [293, 310], there are also indications that other globular clusters did indeed form very early on (see Figure 3.8). We know that globular clusters in the Milky Way span a range of metallicities and a range of properties. Some are metal poor ($Z \sim 10^{-2} Z_\odot$) and are compatible with being simple stellar populations, i.e. containing stars with the same metallicity and age, while others, such as Omega Cen, show a range of properties in their stellar populations [287].

Assuming that the thermonuclear yields computed for Population III stars are correct, globular clusters cannot be second-generation objects, as the Population III

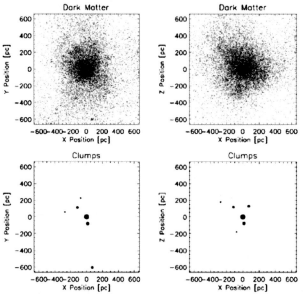

Fig. 3.8 Morphology of a dwarf galaxy halo at $z = 15$ [41] with a box size of 1.3 kpc. The top panels show the dark-matter distribution and the bottom panels indicating dense clumps with the size indicating their mass scale. The left and right panels correspond to two different views of the same system (Reproduced by permission of the AAS).

metal abundance ratios, and in particular the distinctive odd–even pattern predicted for Population III stars are not observed in globular clusters.

Another serious issue is that globular clusters appear not to contain dark matter [175] but it may be difficult to imagine a scenario of isolated formation of a primordial globular cluster that would strip the protoglobular clusters of their dark-matter content [176].

Finally, it is worthwhile highlighting a prediction from the Peebles and Dicke paper, namely that globular clusters if they really formed as an isolated system at high redshift should also exist in isolation outside a galaxy. This prediction is independent of the specific model and is still worth pursuing.

3.5
The First Galaxies

We have seen in Section 3.2.3 that it is already unlikely for a halo with mass of $10^8 M_\odot$ to form primordial stars because it is very likely that the halo will have been polluted by previous generations of stars. The likelihood that more-massive halos can form primordial stars will be even lower.

In order to estimate the brightness of one of the first galaxies, let us consider the mean number of stars simultaneously active in a galaxy of about $3 \times 10^7 M_\odot$ like the one studied by Wise and Abel [298]. This number is obtained by considering the redshift of formation of each star and assigning to each a lifetime of 3 Myr. The resulting curve is shown – after smoothing – in Figure 3.9. The figure shows that there is about one star active at any given time and therefore this object would be very difficult to detect. Assuming that each star has a mass of $300 M_\odot$ the resulting star-formation rate is $10^{-4} M_\odot$ yr^{-1}. Detecting, at a similarly early stage, an object with a $1 M_\odot$ yr^{-1} rate and estimating the halo mass by linear rescaling gives us a required halo mass of around $10^{11} M_\odot$, which would be very rare at $z = 15$ with a density of less than one per Gpc^3 judging from Figure 2.6. This rough estimate gives us an indication of how difficult it is going to be to detect a galaxy made almost exclusively of primordial stars.

On the other hand, it is likely that the star-formation rate will increase significantly as the metallicity increases. Present galaxies have about half of the baryons in stars and have managed to form these stars over about a Hubble time. This extremely simple-minded criterion would suggest that a galaxy forming about $1 M_\odot$ yr^{-1} would have now a stellar mass of $\sim 10^{10} M_\odot$ and sit in a halo of a mass of $2 \times 10^{10}/0.173 \sim 10^{11} M_\odot$. After the first 500 Myr of its life this galaxy would have a stellar mass 20 times smaller and sit in a halo of $5 \times 10^9 M_\odot$. Using the Bruzual–Charlot models [43] we derive for such objects an absolute magnitude $M_{1400} \approx -20$, which should be easily detectable due to very high redshift. Using instead a calibration of luminosity versus mass [60, 279, 284] one finds that galaxies of such mass roughly correspond to $M_{1400} \approx -17$, i.e. to the faintest objects detected in the HUDF [279] at $z \approx 6$. Thus, there is a wide range of predictions for the luminosity of these early galaxies. Based on CDM halo statistics

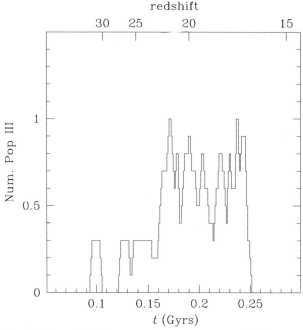

Fig. 3.9 The number of stars simultaneously present in the first 150 million years of the life of a $3 \times 10^7 M_\odot$ halo. This curve has been obtained from the simulation of Wise and Abell [298] after assigning a 3-Myr lifetime to each star and by smoothing the result with a 5-pixel boxcar.

they are predicted to have a surface density of about a few per square arcmin at $z \approx 10$.

3.6
The First Active Galactic Nuclei

Population III stars can leave black hole remnants and these can in turn act as mini-AGNs and even seed more-massive AGNs powering the first active galactic nuclei.

3.6.1
Population III Black Holes

Population III stars with mass in the range ~40 to ~140$M_\odot$ or greater than 260$M_\odot$ may collapse to form black holes without undergoing a pair-instability supernova explosion [113]. This makes it possible for these black holes to be the seeds of the massive black holes in the present-day Universe [159]. However, even these stars create winds because of their strong radiative feedback and these winds are estimated to have velocities around 30 km/s [297].

In Section 3.2.2 we derived the self-shielding of small halos from the Lyman–Werner background and derived the cumulative number density of Population III stars in the presence of a self-regulating background shown in Figure 3.4. The figure shows a density of less than 100 Population III stars per Mpc^3 at redshift $z = 10$. If all of these stars left $100 M_\odot$ black hole remnants this would imply $\approx 10^4 M_\odot$ in black holes per Mpc^{-3}. The local density in black holes is estimated to be $(2\text{--}10) \times 10^5 M_\odot$ [54, 165]. The density we derive is lower and thus leaves room for further mass increase by accretion. It is therefore compatible with the local value.

We expect that black holes will start accreting as soon as gas becomes available. However, we saw in Section 3.3.2 that only halos with mass exceeding a few $10^6 M_\odot$ can contain their circumstellar gas. This can happen either by cooling or by enclosing the heated material in the collapse of a subsequent halo. The median delay for halos of, e.g., $10^6 M_\odot$ to be incorporated in halos of mass $\sim 5 \times 10^6 M_\odot$ able to contain the ionized circumstellar medium (see Table 3.3) is in the range 30–60 Myr for halos collapsing at $z \gtrsim 25$. The delay for the case of cooling is more dependent on the specific configuration and on the density after the initial expansion but if the density remains within a factor of 100 of its initial post-collapse value, the cooling timescale from the cooling terms in (3.24)–(3.27) is $\lesssim 40$ Myr and the same is true for the recombination timescale. Thus, a typical black hole will need to wait $\approx (4\text{--}6) \times 10^7$ yr before beginning gravitational accretion and turning into a mini-quasar. This corresponds to a turning on of accretion at redshifts of 25–30 for black holes formed at redshift 40 or higher.

It is instructive to compare the delay time for accretion to the Salpeter time that is given by the exponential timescale for the growth of a black hole accreting at the Eddington rate. This is obtained by equating the Eddington luminosity given by (2.72) to the accretion luminosity of the black hole:

$$4\pi G M m_p c / \sigma_T = \eta \dot{M} c^2 \tag{3.54}$$

where $\eta \approx 0.1$ is the accretion efficiency. Equation (3.54) is a differential equation for the black hole mass M whose solution is an exponential with timescale given by the Salpeter time:

$$\tau_{Salpeter} = \frac{\eta \sigma_T c}{4\pi G m_p} \simeq 4.52 \times 10^7 \text{ yr} \tag{3.55}$$

Thus, the delay in the black hole growth is of only ~ 1 Salpeter time and it will have only a modest effect on the final black hole mass.

3.6.2
Black-Hole Mergers

As we have seen, a black hole remnant cannot begin accreting immediately as the circumstellar gas is hot and ionized. Even after a delay of a few 10^7 years accretion can be hampered by other factors. If halos containing black holes merge with one another, one could end up in the situation where a third black hole is added to a black hole binary. The subsequent evolution of the system could lead to ejection of one of the black holes. Ejection could also be caused by the gravitational recoil as gravitational waves emitted by a rotating black hole pair could carry away linear momentum [289]. Even without ejections the interaction of multiple black holes would most likely disrupt accretion until the black holes merge and would lead to a slower growth in mass than otherwise possible.

We can estimate the probability of a merger from the cumulative number of Population III stars shown in Figure 3.4. The number density reaches a value of about 1 Mpc^{-3} at $z \sim 30$ and then grows slowly, remaining under 100 Mpc^{-3} at $z \sim 10$. A cubic comoving Mpc^3 contains about $3.5 \times 10^{10} M_\odot$ of matter. Thus, halos forming population III stars represent a mass fraction of only $\sim 3 \times 10^{-5}$ at redshift 30 and $\sim 3 \times 10^{-3}$ at redshift 10. Even considering the high bias of these early halos, the probability that two such halos merge at redshift 30 is likely to be essentially negligible, while clustering of these halos might make it more likely at redshift 10 despite their low mass fraction. A more precise estimate would need to include a detailed merger model. In a simulation of the formation of a dwarf galaxy at high-z [102], some of the 10 or so Population III formed before the galaxy assembles, spend more than 10^8 years before merging with other halos containing Population III remnants, having thus had the chance to increase in mass by a factor of ~10, corresponding to 2–3 Salpeter times. Clearly, a higher-mass black hole would have better chances to survive, without being ejected, the merging with a lower-mass one.

3.6.3
The Highest-Redshift QSOs

It is possible to connect the first Population III stars in a large volume with the largest overdensities that we think are associated to bright QSOs at high redshift like, e.g., the QSOs at $z = 6$ identified by the Sloan Digital Sky Survey (SDSS). We have seen in Section 3.6.1 that the start of accretion for a Population III black hole remnant is delayed by about one Salpeter time compared to the ~ 20 Salpeter times between redshift 50 when the first Population III stars in a Gpc^3 volume form and redshift 6 when the highest redshift bright QSOs detected by SDSS have been found [80]. The 19 Salpeter times that one is left with would enable a $100 M_\odot$ seed black hole remnant to grow to a mass $> 10^{10} M_\odot$. Clearly, not all black holes will be able to accrete at the Eddington rate throughout this time but it can be shown that there are thousands of Population III stars formed at high-z in the volume occupied by a bright SDSS QSO [278] and requiring that sufficient gas is available for accretion still leaves a few able to grow to large masses. Moreover, these seed black holes are sufficiently rare at high redshift that merging with other objects will take place only after they have increased significantly their mass. Thus, Eddington accretion in gas-rich halos from seeds provided by Population III black hole remnants remains a viable scenario for explaining the SDSS QSOs. The least palatable aspect of this scenario is that the fast Eddington growth necessary to make the bright redshift 6 QSOs would need to slow down significantly around redshift 6 in order to produce the observed evolution of lower-redshift QSOs.

3.6.4
Direct Collapse to Black Holes

In order to bypass the difficulties in starting off efficient accretion on a Population III remnant, the disruptive effect of mergers, and a fine-tuned duration of the Eddington accretion phase, a different scenario has been proposed to form massive black holes at high redshift, namely the direct collapse to black holes in a $\sim 10^8 M_\odot$ metal-free halo [19]. On the basis of our calculation of pollution it is clear that only a relatively small fraction of these massive halos will survive with pristine gas. On the other hand, even this small fraction could be sufficient to form the observed high-redshift QSOs. The advantage of this scenario is that the black hole growth would not need to be at the Eddington rate at all times and thus one might obtain a less dramatic evolution of the black hole mass function at redshift greater than 6, more in line with what is observed at redshift below 6. A more detailed analysis will need to be carried out before we can assess whether this is a viable scenario.

3.7
Low-Metallicity HII Regions

In order to distinguish a zero-metallicity HII region from one with very low, but nonzero, metallicity we need to compute the properties of the latter objects. We will refer here to an analysis by Panagia and coworkers carried out for metallicity down to $10^{-6} Z_\odot$ [203] using a combination of Cloudy90 [82] and analytical calculations. In Figure 2.15 the synthetic spectrum of an HII region with metallicity $10^{-3} Z_\odot$ is compared to that of an object with zero metallicity. The two are very similar, except for a few weak metal lines with the strongest one being the [OIII]λ5007 line. In Figure 3.10 we show the Lyman α EWs for individual HII regions ionized by stars with different masses and metallicities. Values of EW in excess of 1000 are possible already for objects with metallicity $\sim 10^{-3} Z_\odot$.

Even though the metal lines at these metallicities are weak, some of them can be used as metallicity tracers. In Figure 3.11 the ratio of the intensity of [OIII]λ5007 to Hβ is plotted for a range of stellar temperatures and metallicities. It is immediately apparent that for $Z < 10^{-1} Z_\odot$ this line ratio traces metallicity linearly (for each individual star.) Our reference value $Z = 10^{-3}$ corresponds to a ratio [OIII]/Hβ = 0.1. The weak dependence on temperature ensures that this ratio remains an excellent indicator of metallicity also when one considers the integrated signal from a population with a range of stellar masses.

Another difference between zero-metallicity and low-metallicity HII regions lies in the possibility that the latter may contain dust. If dust can form at low metallicity

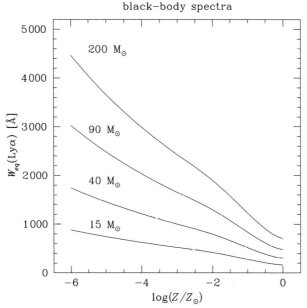

Fig. 3.10 Lyα equivalent width (in Å) as a function of metallicity and stellar mass calculated approximating the stellar atmosphere with a blackbody. For metallicity below $10^{-3} Z_\odot$ and stellar masses above $40 M_\odot$ the Lyα EW exceeds 1000 Å.

as well as in the local Universe, for a $Z = 10^{-3} Z_\odot$ HII region dust may absorb up to 30% of the Lyman α line, resulting in roughly 15% of the energy being emitted in the far IR [203].

3.8
Numerical Techniques and Their Limitations

Numerical simulations have become part of the tool set of modern cosmology and many of the results described in this chapter rely on the results of sophisticated simulations. In the following we will review a few of the most important techniques for numerical simulations and we will also discuss their limitations.

3.8.1
Collisionless Dynamics

Stars and dark matter in galaxies behave as collisionless particles and their dynamical evolution is dictated only by gravity. The dynamics of a system evolving solely under the force of gravity is described by a system of equations that, for each parti-

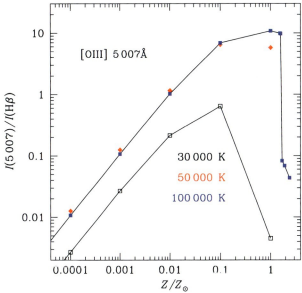

Fig. 3.11 [OIII]λ5007/Hβ ratio as a function of metallicity for HII regions ionized by stars with different effective temperatures.: 30 000 K (open squares and lower curve), 50 000 K (diamonds), 100 000 K (filled squares). Above 5×10^4 K the [OIII]λ5007/Hβ ratio does not depend on the stellar temperature. At metallicities below $10^{-2} Z_\odot$ it is roughly a linear function of metallicity.

cle i, are of the form:

$$\ddot{x}_i = -\sum_{j, i \neq j}(x_j - x_i)\frac{G m_i m_j}{|x_j - x_i|^3} \tag{3.56}$$

where m_i is the mass of the ith particle. For a system that can be considered collisionless, the evolution is dominated by the mean gravitational field derived from Poisson's equation:

$$\nabla^2 \Phi = 4\pi G \varrho \tag{3.57}$$

where the smooth density ϱ is derived from the particle distribution. The equations of motion are in this case simply:

$$\ddot{x}_i = -\nabla \Phi(x_i) \tag{3.58}$$

On scales smaller than galaxies, dark matter (probably) continues to behave as collisionless particles but collisionality plays a role in, e.g., the evolution of globular clusters. Clearly collisionality in this context does not refer to physical collisions between stars but rather to the fact that close encounters become important in determining trajectories of individual stars and therefore that a mean-field approximation for the gravitational field experienced by any single star is no longer valid.

The relevant cross section for star encounters is not the physical cross section but the gravitational cross section:

$$\sigma_{grav} \approx \pi \left(\frac{GM_*}{c_s^2}\right)^2 \qquad (3.59)$$

where c_s^2 is the stellar velocity dispersion squared. A collision rate can be derived from the cross section as: $\tau^{-1} = n\sigma_{grav}c_s \propto c_s^{-1}$, where n is the number density. Thus, for a stellar system – all else being equal – a higher velocity dispersion implies a lower collision rate. In contrast, for a system with a geometric cross section, higher velocity dispersion implies higher collision rate. The degree of collisionality of a system can be measured by the ratio of the collisional relaxation timescale τ_{rel} over a dynamical timescale provided, for instance by the free fall time τ_{ff}. It is possible to show that for a self-gravitating system of N particles interacting only by gravity one has [26, 50]:

$$\frac{\tau_{rel}}{\tau_{ff}} \simeq \frac{0.1N}{\ln N} \qquad (3.60)$$

Equation (3.60) implies that for a stellar system with a small number of particle collisions are important but confirms that for galaxies they can be ignored. Below we will describe the two major types of dissipationless codes used in cosmological applications. For a more indepth review of N-body techniques the reader should consult the review by Trenti and Hut [277].

3.8.2
Collisionless Dynamics: Particle-Mesh Codes

Any code to study the dynamical evolution of a collisionless system will be based on a Poisson solver computing the gravitational force at any desired location and a solver for the equation of the motion of each particle. An approach to do so is represented by the so-called particle-mesh codes.

Various implementations of particle-mesh codes have been for decades the primary tools for numerical cosmology. These codes are broadly based on the idea of adopting a mean-field approach in which the positions of individual particles are used to derive a mean gravitational force field applied to the particles (PM code). In order to bypass the limitation to the spatial resolution imposed by the grid size, the contribution from particles within a cell is generally subtracted and the direct particle–particle forces within a cell are added back so as to include small-scale variations of gravitational field that would occur on scales much smaller than the grid size [76] (P3M codes). A further version of these codes allows for grid refinements either predefined or adaptive so as to increase the effective resolution of the code where it is most needed [63] (AP3M codes).

PM codes generally rely on the use of fast Fourier transforms to solve Poisson's equation and, in a cosmological context, are written in comoving coordinates and rely on periodic boundary conditions.

3.8.3
Collisionless Dynamics: Treecodes

Tree codes are based on the idea of expanding in spherical harmonics the gravitational force acting on a particle retaining only a few of the lowest orders, often just the force and the quadrupole component [13, 114]. In practice, a 'tree' is built pointing to each particle. In the language of computer science a 'tree' is a hierarchical structure with branches connecting nodes. Each node v_i contains a group of particles or a single particle. In this type of code, each node connects to a single higher-level node, v_{i-1}, grouping several particles including those in the node under consideration. The node v_i also points to several lower-level nodes, $v_{i+1,j}$, containing, either singly or in small groups, the particles of which the node is composed. The root node v_0 is the highest-level node and points, indirectly, to the whole system. When deriving the force acting on a particle by a group of particles on a distant branch one generally applies an angular criterion, namely if the group of particles subtends an angle smaller than some threshold opening angle their contribution to the force is added as a single superparticle in the spherical-harmonics approximation. When the angle subtended exceeds the threshold, the particles in the given branch cannot be considered together and one considers the sub-branches of which the branch being analyzed is composed.

An interesting feature of this type of code is that the choice of the opening angle determines how close the potential used to evolve the system is to the gravitational potential (through the use of truncated spherical harmonics) but the integration can be apparently accurate even when it is not. Thus, a simple energy-conservation test cannot be used to assess the validity of an integration. Since the integration time increases with decreasing opening angle, in practice the range of acceptable and practical opening angles is limited.

3.8.4
Gas Dynamics

The dynamics of a gaseous system are described by a system of differential equations. From the point of view of statistical mechanics fluid equations are just a series of moments of the equations governing the distribution function of the system. The equation for each moment involves moments of a higher order and in general they can be closed only by assuming some additional, physics-driven equation to close the system. This is often done by assuming an equation of state. A minimal system of equations includes the continuity equation expressing the conservation of mass:

$$\frac{\partial \varrho}{\partial t} + \nabla \cdot (\varrho v) = 0 \qquad (3.61)$$

where v is the velocity field, and Euler's equation that represents the fluid analog of the equation of the motion and is written as:

$$\varrho \frac{\partial v}{\partial t} + \varrho (v \cdot \nabla) v = -\nabla P - \varrho \nabla \Phi \tag{3.62}$$

where we have assumed that the fluid is in a gravitational field and subject to pressure forces P. The closure of this system can be obtained by adopting an equation connecting pressure to temperature and density. Two popular assumptions are the polytropic equation of state:

$$P = K \varrho^\gamma \tag{3.63}$$

where K and γ are two constants, or the perfect gas equation of state:

$$P = \frac{\varrho k T}{\bar{m}} \tag{3.64}$$

where $\bar{m}$ is the mean molecular mass of the gas. In more complex cases one will have a multifluid system with subsystems connected by chemical evolution equations or even systems with heat sources or sinks requiring also an energy equation.

3.8.5
Gas Dynamics: Smooth Particle Hydrodynamics

Smooth particle hydrodynamics (SPH) codes are the most popular implementation of Lagrangian codes that follow individual fluid elements, decoupling the motion of the fluid elements from their internal evolution in terms of temperature, pressure, density. In SPH codes fluid-element particles can be treated similarly to mass particles in collisionless codes for the purposes of gravitational evolution. Gaseous forces require the calculation of pressure gradients and these are computed through the so-called SPH-kernel allowing estimation of fluid derivatives starting from a sampling of fluid quantities by particles at random spatial locations. SPH codes are easy to use and relatively robust. However, they have limited dynamic range especially when dealing with strong shocks. A popular, public domain, SPH code for cosmological applications is Gadget [253].

3.8.6
Gas Dynamics: Eulerian Codes

In Eulerian codes the coordinates system does not follow the fluid elements. This generally complicates matters but may be a decisive advantage in situations where the positions of, e.g., a shock front is known in advance. Integration of the fluid equations in a Eulerian forward time explicit form tends to exhibit numerical instabilities so that some of the most successful Eulerian codes are implicit, i.e. the quantities at time step t_{i+1} are derived by solving an implicit equation or are based on a multistep solution where different operators are applied in a sequence [267]. These techniques tend to guarantee numerical stability. A popular, publicly available, Eulerian code for cosmological applications is the adaptive mesh refinement code Enzo [190].

3.8.7
Improving Resolution Through Mesh Refinement

Adaptive codes or treecodes can simulate dynamics on small scales. However, in the cosmological context adaptivity does not guarantee that the correct dynamics is reproduced. The issue here is that of the initial conditions. If initial conditions for a simulation are generated without including power below some small mass scale m, evolution at scales smaller than m cannot be trusted. In many applications having to do with the first stars, one is interested in studying rare objects of small mass over a large volume. This quickly becomes very demanding in terms of computing resources unless one can make use of refinement techniques. These techniques enable one to address similar problems but should be used beign fully aware of their limitations [95]. In particular, one has to remember that the descendants of rare objects at high redshift are statistically more and more common with decreasing redshift [280].

3.8.8
Radiative Transfer

Radiative transfer is a complex problem and many existing codes rely on ray tracing often on scales larger than those used for dynamical evolution. Many of the experts on this field joined a comparison project of the various codes [117]. The comparison for static density fields was carried out using 11 independent codes and 5 test cases and shows good, if not perfect, agreement. The study represents a benchmark for future developments in this area.

3.9
Hints for Further Study

- Local dwarf galaxies often have a central star cluster. Assuming the clusters formed before the galaxy and adopting a Salpeter initial mass function and Bruzual–Charlot models [43], compute a lower limit to the metallicity of these dwarf galaxies due to their central cluster [143].
- Considering the metallicity of the metal poor dwarf galaxy I Zw 18 [7, 302] how many Population III stars would have been needed to enrich it? how many 30 Msun Population II stars?
- A few stars in the Milky Way have been found with iron metallicity below 10^{-5} the solar level [55, 56, 88]. Could the gas from which these stars formed be enriched by a single Population III star? allowing if necessary also contamination by intermediate generation stars, what is the minimum gas mass that was enriched by a Population III star?

4
Cosmic Reionization

4.1
Overview

In the standard cold dark matter (CDM) cosmology galaxies are assembled through hierarchical merging of building blocks with smaller mass. As we have seen in Chapters 2 and 3 (see also [2, 3, 62, 104, 106, 198]), the first such building blocks with primordial chemical composition and able to cool gas, have $M \gtrsim 10^5 M_\odot$ and form in ΛCDM models at $z \gtrsim 30$.

The energy to ionize hydrogen (13.6 eV) is much lower than that to ionize once (24.6 eV) or twice (54.4 eV) helium. Thus, it is generally believed that stars are responsible for the reionization of hydrogen [154] at $z \simeq 6-20$ [53, 101, 105] while the harder UV spectrum of AGN is responsible for the reionization of helium [111, 119] at lower redshift when the AGN comoving number density peaks.

The following simple calculation shows that nuclear processing of only a very small fraction of baryons would be sufficient to reionize the Universe [141]. Fusion of hydrogen releases 7 MeV per proton, but only 13.6 eV are needed to ionize hydrogen. Thus, a fraction $\sim 2 \times 10^{-6}$ of hydrogen undergoing fusion is energetically sufficient to reionize all hydrogen. In practice, the minimum fraction will be larger by a factor of ten to one hundred because not all energy is released in the form of ionizing photons, not all ionizing photons have the minimum ionization energy, not all ionizing photons successfully escape from their sources into the IGM, and recombinations increase the required number of ionizing photons. All these factors depend on the details of the process and have large uncertainties. By assuming that the minimum fraction is 30 times the value given above, i.e. 6×10^{-5} and a yield of $\sim 0.3 M_\odot$ of metals produced per $M_\odot$ of hydrogen processed in a massive star [113, 187], one can estimate that the minimum mean metallicity of the Universe at reionization was $\sim 10^{-3} Z_\odot$. These arguments will be quantified in more detail in the following sections.

If the power source for reionization is not nuclear fusion but rather gravitational accretion onto massive black holes, a smaller fraction of material needs to be processed thanks to the higher efficiency of gravitational accretion. Clearly this scenario does not place any constraint on the metallicity of the Universe at reionization and leaves unanswered the question of the origin of the black holes (primordial or

From First Light to Reionization. Massimo Stiavelli
Copyright © 2009 WILEY-VCH Verlag GmbH & Co. KGaA, Weinheim
ISBN: 978-3-527-40705-7

stellar). Even if reionization is caused by stellar UV radiation, it is natural to expect that some fraction of these stars will leave black holes as remnants [159]. Thus, in both scenarios, we expect that some seed black holes will be present at the end of reionization, with interesting implications on the formation of AGN and galaxies [243, 263]. It is worth noting that both Population III stars and their black hole remnants accreting in a gas-rich halo have comparable luminosities close to the Eddington value.

Even though we often refer to reionization as if it was a sudden transition, the time elapsed between the epochs when 10% and 90% of hydrogen was reionized can last a significant fraction of the age of the Universe at reionization: thus, the dark ages may end with an extended twilight. Sufficiently dense neutral-hydrogen clouds can self-shield and will not be easily ionized by a diffuse background. Indeed, discrete neutral-hydrogen clouds are detected at all redshifts from their Lyman α seen in absorption in the spectra of high-redshift QSOs. Thus, a more useful definition for the end of the reionization of hydrogen is not the complete ionization but percolation, i.e. the time when the ionized hydrogen bubbles touch, changing the topology of the ionized medium. Percolation is an especially useful concept if reionization is done by stars. AGN would have a hard spectrum extending to the X-rays and the ionization cross section of X-rays is lower so that, in the case of reionization by AGN, ionized bubbles would have less sharp boundaries and there could also be partly ionized medium in between the fully ionized bubbles.

In practice, percolation is hard to measure and one will have to resort to less-direct indicators. Inhomogeneities along the line of sight may create a dispersion in optical depth shortwards of Lyman α. Moreover, only a very low residual fraction of neutral hydrogen is needed to produce a Gunn–Peterson trough in the spectra of high-redshift quasars. To illustrate this we give in Figure 4.1, as a function of redshift, the fraction of neutral hydrogen required to have a Gunn–Peterson optical depth of one or ten in a uniform IGM [16, 103]. It should be noted that in analogy to the proximity effect in QSOs [96, 100, 144, 174, 204] the opacity near Lyman α would be modified in the neighborhood of ionizing sources [168]. In order to obviate the problem of the extreme sensitivity of the Gunn–Peterson trough to even a very small residual fraction of neutral hydrogen, methods are being considered that rely on the study of the microwave background polarization [125] or on 21 cm diagnostics or other techniques (see Chapter 5).

In the following sections, we will derive the observable properties of the sources responsible for the reionization of hydrogen starting from the simplest assumptions and considering increasingly more realistic cases.

4.2
The Properties of the Sources of Reionization

Here, we will consider the properties of ionizing sources in the simple cases of a homogeneous Universe composed only of hydrogen or of a mixture of hydrogen

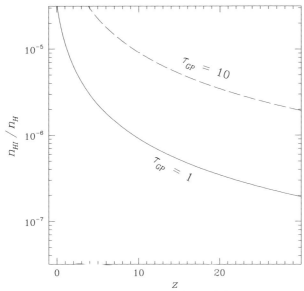

Fig. 4.1 The figure shows as a function of redshift the hydrogen ionization fraction required to have a Gunn–Peterson optical depth of one (solid line) or ten (dashed line).

and helium. The methods outlined in this section are based on those developed by Stiavelli and coworkers [264].

4.2.1
The Surface Brightness of Reionization Sources

There is a simple connection between the comoving density of hydrogen and the mean surface brightness J_ν of the reionization sources in their nonionizing continuum [170]. Below, we derive this dependence. Neglecting the absorption of non-ionizing photons in the IGM, the surface brightness of a class of sources is simply related to their volume emissivity $E(\nu)$, i.e. the power radiated per unit frequency per unit comoving volume, by

$$J_\nu = \frac{c}{4\pi} \int_{t_1}^{t_2} E([1+z(t)]\nu) \, dt \tag{4.1}$$

where t_1 indicates the cosmic time when the sources turn on and t_2 the time when reionization is complete. The production rate of ionizing photons per unit comoving volume $\dot{n}_c$ is related to the comoving volume emissivity by:

$$\dot{n}_c = \int_{\nu_c}^{\infty} \frac{E(\nu)}{h\nu} \, d\nu \tag{4.2}$$

where h is Planck's constant and $\nu_c = 2.467 \times 10^{15}$ Hz is the frequency of photons at the hydrogen ionization threshold, known as the Lyman limit. Here, and in the following, we denote with the subscript c quantities referring to Lyman continuum photons.

For a given spectral energy distribution (SED) of reionization sources, there is a direct proportionality relation between the production rate of ionizing photons $\dot{n}_c$ and the emissivity $E(\nu_0)$ at some reference frequency ν_0. With this in mind, we define the proportionality coefficient as:

$$A(\nu_0) \equiv \frac{E(\nu_0)}{\dot{n}_c} \tag{4.3}$$

Note that $A(\nu_0)$ has the dimensions of an action and is of the same order of magnitude as Planck's constant h. In the following, we choose ν_0 to be a fixed frequency in the observer frame in the nonionizing continuum of the sources as only the nonionizing continuum will in general be accessible to observations.

In order to ionize hydrogen completely, the total number of ionizing photons per comoving volume must exceed the comoving number density of hydrogen n_H by some factor B that depends on the details of the recombination process and on the helium fraction and is defined as:

$$B \equiv \frac{1}{n_H} \int_{t_1}^{t_2} \dot{n}_c \, dt \tag{4.4}$$

Combining (4.1), (4.3), and (4.4), we obtain

$$J_\nu = \frac{c \bar{A}(\nu) B}{4\pi} n_H \tag{4.5}$$

where $\bar{A}(\nu)$ denotes the average of $A(\nu_0)$ weighted by the production rate of ionizing photons, namely

$$\bar{A}(\nu) \equiv \frac{\int_{t_1}^{t_2} A([1+z(t)]\nu) \dot{n}_c \, dt}{\int_{t_1}^{t_2} \dot{n}_c \, dt} \tag{4.6}$$

Equation (4.5) establishes a simple, direct relation between the mean surface brightness of reionization sources and the comoving density of hydrogen atoms. In the following, for simplicity, we will neglect possible variations in $\dot{n}_c$. We can approximate $1 + z \propto t^{-2/3}$ and write:

$$\dot{n}_c \equiv \frac{dn_c}{dt} = \frac{dn_c}{dz} \frac{dz}{dt} \propto \frac{dn_c}{dz} t^{-4/3} \tag{4.7}$$

Replacing again t with $1 + z$ we find:

$$\frac{dn_c}{dz} \propto \dot{n}_c (1+z)^{-2} \tag{4.8}$$

From (4.8) we see that the assumption of a constant $\dot{n}_c$ is equivalent to $\frac{dn_c}{dz}$ decreasing with redshift because the time element is compressed at high z. It is possible to repeat the whole derivation from (4.1) to (4.6) using redshift instead of time as the independent variable. In order to do so, the approximation of $\dot{n}_c$ constant is replaced by $d\dot{n}_c/dz$ = constant that implies $\dot{n}_c$ increasing with redshift, which appears unrealistic.

When $\dot{n}_c$ is constant, $\bar{A}(\nu)$ becomes a time average:

$$\bar{A}(\nu) \equiv \frac{1}{t_2 - t_1} \int_{t_1}^{t_2} A([1 + z(t)]\nu) \, dt \qquad (4.9)$$

The volume emissivity $E(\nu)$ for a population of sources is given in terms of their luminosity function per comoving volume, $\Phi(L_\nu)$, as

$$E(\nu) = \int_0^\infty \Phi(L_\nu) L_\nu \, dL_\nu \qquad (4.10)$$

Equations (4.1) and (4.10) establish the connection between the mean surface brightness of a population of sources and their luminosity function. In the case of identical sources the luminosity function is simply a delta-function. The following subsections will be devoted to estimating the values of $\bar{A}(\nu)$ and B.

4.2.2
Reionization in a Hydrogen-Only IGM

Due to recombinations, the minimum number of ionizing photons per comoving volume needed to reionize hydrogen exceeds the number of hydrogen atoms ($B > 1$). The recombination rate is proportional to the square of the local (noncomoving) density. Thus, inhomogeneities are going to play a major role in determining the value of B by acting as sinks of ionizing UV photons. We can account for IGM inhomogeneities by including a clumping factor $C_\varrho \equiv \langle n_p^2 \rangle / \langle n_p \rangle^2 = \langle n_e^2 \rangle / \langle n_e \rangle^2$ in front of the recombination term valid for a homogeneous Universe (which corresponds to $C_\varrho = 1$).

The equation for the ionized fraction $x \equiv n_p/n_H$ can be derived from (2.3) following the steps leading to (2.6) but replacing the ionization term in β_H with $\dot{n}_c$ and including the clumping factor C_ϱ to obtain:

$$\frac{dx}{dt} = \frac{\dot{n}_c}{n_H} - C_\varrho \alpha_B x (1 + z(t))^3 n_e \qquad (4.11)$$

where n_e is the comoving density of electrons and $\alpha_B = 1.63 \times 10^{-13}$ cm^3 s^{-1} is the recombination rate for an electron temperature of $T_e = 2 \times 10^4$ K obtained from (2.11). For the same temperature, Stiavelli et al. [264] use $\alpha_B = 1.4 \times 10^{-13}$ cm^3 s^{-1}, while Osterbrock and Ferland [197] give $\alpha_H = 1.43 \times 10^{-13}$ cm^3 s^{-1}. The benefit of using the value of (2.11) is that by doing so it is easy to consider other values of the temperature. The value of the electron temperature of $T_e = 2 \times 10^4$ K is appropriate for primordial HII regions [65, 67, 282] but the electron temperature at reionization is probably higher (see Section 4.7) and the recombination coefficient will be even lower, so that recombinations will be less important. The factor $(1 + z)^3$ is due to the fact that recombinations depend on the proper (not comoving) density of electrons.

For the pure-hydrogen composition adopted in this section we have $n_e = n_p$. Solving (4.11) and requiring $x(t_2) = 1$, we can determine $\dot{n}_c$ as a function of t_1 and t_2.

The parameter B of (4.4) is the ratio of the actual number density of ionizing photons to the number density of hydrogen atoms. We define B_H as the value of B in the pure-hydrogen case. In the absence of recombinations (i.e. in the limit of a very low density) and in the pure-hydrogen case, the required comoving number density of ionizing photons, $n_{c,\text{noHe}} = \dot{n}_c(t_2 - t_1)$, equals the comoving number density of hydrogen, $n_{c,\text{noHe}} = n_H$ and therefore $B_H = 1$. When recombinations are considered in the homogeneous case ($C_\varrho = 1$), we find $B_H \leq 1.03$ or ≤ 1.08 for $z_2 = z(t_2) \lesssim 10$, and $\Delta z \equiv z(t_1) - z(t_2) = 1$ or 3, respectively. We also find that if reionization occurs at $z_2 = 6$, $B_H \leq 1.3$ for $\Delta z \leq 50$. Thus, in a homogeneous ($C_\varrho = 1$) pure-hydrogen IGM, recombinations are only a minor correction.

4.2.3
Reionization in a Hydrogen–Helium IGM

More realistically, the chemical composition of the universe at reionization includes both hydrogen and helium, with mass fractions $X = 0.74$ and $Y = 0.26$ [200]. Thus, the mean molecular weight is given by $\bar{m} = (m_H Y/m_{He} + X)^{-1} m_H \simeq 1.24 m_H$, where m_H and m_{He} are the masses of hydrogen and helium atoms, respectively. We find that the number density of hydrogen atoms (including ions) per comoving volume, n_H is 1.8×10^{-7} cm^{-3} or 5.4×10^{66} Mpc^{-3}.

Photons with energies above 24.6 eV can ionize helium as well as hydrogen. The cross section for helium ionization at the threshold is higher than that of hydrogen by about a factor 7 but the number density of He is lower by a factor of about 12, so that photons at these energies may be absorbed by, and ionize, either atoms. In the following we will assume that helium absorbs as many photons as possible in proportion to its abundance relative to hydrogen, i.e. about 8% of the total budget of ionizing photons. Photons with energy above 54.4 eV can ionize helium twice. Whether helium is ionized or not and how many times, depends on the hardness of the UV spectrum. As an example, a blackbody at $T = 10^5$ K emits 50% of its ionizing photons at energies greater than 24.6 eV. In this case, all the helium will be ionized once, and about 10% will be twice ionized. Therefore, we have $n_e \simeq 1.09 n_p$, and the required minimum number of ionizing photons will be higher by 9% relative to the pure-hydrogen case. Note that if the spectral energy distribution is hard enough to double-ionize all helium, n_e increases to $1.16 n_p$. This is unlikely to happen for stellar sources, but it might for active galactic nuclei (AGN).

Helium ionization has an additional, indirect effect on the required number of ionizing photons because the 9% increase in the electron number density makes recombinations slightly more effective. This can be estimated by integrating (4.11) for $n_e \simeq 1.09 n_p$, from which we obtain

$$B \simeq 1.09[1 + 1.09(B_H - 1)] \tag{4.12}$$

In general, we find that primordial helium recombinations increase B by less than an additional 1% for $z_2 \leq 10$ and $\Delta z \leq 3$ and by less than 2% for reionization at $z_2 = 6$ and $\Delta z \leq 50$. Thus, recombinations remain a minor correction in the homogeneous ($C_\varrho = 1$) case, even for a hydrogen–helium IGM.

4.2.4
Results for a Homogeneous IGM

So far, the only dependence of our calculations on the type of ionizing sources has been in the assumption that all helium is ionized once. Indeed, even if this assumption were not verified, our results would change by only 10% or less. In order to derive the mean surface brightness of sources from their production of ionizing photons, we need to assume a specific SED and compute the quantity $\bar{A}(\nu)$. While we do not expect reionization to be completed by stars with primordial metallicity, it is useful to consider Population III stars as possible sources responsible for reionization of hydrogen for two reasons. The first is that Population III stars will provide us with the limiting case of very efficient ionizers, the second is that they can also be a proxy for a top-heavy population of very low metallicity stars. As we have seen in Chapter 2, Population III stars are characterized by virtually zero metallicities and consequently very high effective temperatures [39, 49, 163, 281]. Here, we will describe them as blackbodies with effective temperatures of 10^5 K. Note that high temperatures around 10^5 K are reached even if the mass function does not extend beyond $100 M_\odot$ [12, 231, 281]. Population III stars thus provide us with the minimum mean surface brightness needed for the case of reionization by stars. In fact, for a fixed flux of ionizing photons, cooler star are brighter in the visible and in the nonionizing UV continuum. One can define the efficiency of an ionization source as proportional to the ratio of the number of ionizing photons to the total energy output. Cooler stars emit a smaller fraction of their radiation in the form of ionizing photons and are therefore less-efficient ionizers. As an example, for equal ionizing fluxes, a stellar population with $T_\star = 5 \times 10^4$ K is a factor ~4.5 brighter at $\lambda = 1400$ Å than a stellar population with $T_\star = 10^5$ K.

A similar comparison can be made for typical AGN. When considering only primary ionizations, we find that for typical power-law SED $\propto \nu^{-0.5}$ [210], AGN are less efficient ionizers than Population III stars. Indeed, considering the difference in SED and also the full ionization of helium, the minimum surface brightness of reionization by QSOs is brighter than that by Population III stars by 0.776 magnitudes. However, the excess energy of AGNs extends to the X-rays and secondary ionizations by the photoelectrons become possible [240]. Using the results derived by Shull [240] we find that the number of secondary ionizations depends on the ionized fraction and for low ionized fractions ($x \leq 10^{-4}$) it can be 4–10 times larger than the number of primary ionizations (for cutoffs between 1 and 10 keV). The corresponding number for Population III stars would only be 22% since their spectrum is less hard than that of AGNs. When averaging over all values of the ionized fraction the number of secondary ionizations for each primary ionization for AGNs ranges between 0.7 and 2.64 depending on the energy cutoff. A more specific estimate would require more detailed modeling to simulate the various forms of energy loss of the energetic electrons produced by the primary ionizations. The equivalent correction is 2% for Population III stars. When secondary ionizations are taken into account AGNs range from being slightly less efficient to being more efficient than Population III stars. Their minimum surface brightness range between 0.2 mag brighter and 0.627 mag fainter than that of Population III stars.

Once the SED is specified, we can determine the value of $\bar{A}$ from (4.3) and (4.9). In integrating (4.9), the emissivity can be assumed to be zero below the rest-frame wavelength of 1216 Å since during the reionization era these photons will be absorbed by intervening neutral hydrogen. In magnitude units, the surface brightness per square arcmin is given by $\mu_{AB} = -2.5 \log J_\nu + 8.9$, when J_ν is expressed in Jy arcmin^{-2}. Typical values are around $AB = 29$ mag arcmin^{-2}.

In Table 4.1, we give the minimum surface brightness of reionization sources for a variety of models with terminal redshifts of reionization $z_2 = 6$ and different duration Δz. Three values of surface brightness are given in the table for each model, corresponding to the rest-frame 1400 Å wavelength observed at the terminal redshift of reionization z_2 but computed for Population III stars with an effective temperature of 10^5 K, Population II stars with top-heavy IMF, low metallicity and an effective temperature of 5×10^4 K and a solar metallicity population with normal IMF. In addition, we also give the number of photons per ionization B and the Thomson opacity τ_T, integrated to the beginning of the reionization process, and given by (2.13).

All results in Table 4.1 include the effects of helium ionization (by massive hot stars) and recombinations. In the homogeneous case ($C_\varrho = 1$), the minimum surface brightness (in flux, maximum in magnitude) is essentially independent of the terminal redshift of reionization (within 0.1 magnitudes for $\Delta z = 1$) but does depend on Δz. This is why in the table we have focused on a single value of z_2. It is clear that the minimum surface brightness required in the case of a solar population is very different from that obtained for Population III stars. The galaxies responsible for reionization are easier to detect if their metallicity is higher.

The value of the Thompson opacity is a strong function of Δz. In the case of extended reionization, $\Delta z = 18$, we obtain values of $\tau_T \simeq 0.09$ compatible with the 5-yr WMAP data ($\tau_T = 0.084 \pm 0.016$ [115, 178]). The empirical 2σ limit, $\tau_T \geq 0.05$, implies $\Delta z \geq 1$. In the figures, we will focus on the $\Delta z = 1$ case since it is close to the minimum ionizing photon requirement $B = 1$ (obtained in the absence of recombinations), while remaining marginally compatible with the WMAP results.

The numbers given in Table 4.1 can be used directly to obtain the luminosity of the sources of ionizing photons on the assumption that they are a population of identical objects with a specified surface number density. Indeed, the numbers in the table represent the apparent magnitudes of individual objects if their surface number density is 1 arcmin^{-2}. A simple rescaling yields the apparent magnitude m for different assumed surface densities $\mathcal{N}$ namely

$$\log \mathcal{N} = 0.4(m - \mu_{AB}) \tag{4.13}$$

Tab. 4.1 Properties of a series of models completing reionization at $z = 6$. As a function of the duration of reionization Δz (col 1) and clumpiness C_ϱ (col 2), we give the effective number of photons per reionization B (col 3), the Thompson optical depth τ_T (col 4) and the surface brightness per square arcmin for reionization by 100 K sources (col 5) for 50 K sources (col 6) and for solar metallicity sources (col 7).

Δz	C_ϱ	B	τ_T	$\langle \mu_{1400}^{100\,K} \rangle$	$\langle \mu_{1400}^{50\,K} \rangle$	$\langle \mu_{1400}^{solar} \rangle$
1.0	1.0	1.116	0.045	28.94	27.30	25.30
1.0	5.0	1.231	0.045	28.83	27.19	25.19
1.0	30.0	2.362	0.046	28.13	26.48	24.49
4.0	1.0	1.183	0.057	28.88	27.23	25.24
4.0	5.0	1.662	0.060	28.51	26.86	24.87
4.0	30.0	6.293	0.067	27.06	25.42	23.42
18.0	1.0	1.328	0.090	28.75	27.11	25.11
18.0	5.0	2.607	0.101	28.02	26.38	24.38
18.0	30.0	11.07	0.108	26.45	24.81	22.81
18.0	GO	1.266	0.059	28.80	27.16	25.16

where μ_{AB} is the mean surface brightness given in Table 4.1. This relationship for $z_2 = 6$, $\Delta z = 1$, and an observer wavelength $\lambda = 1400(1 + z_2)$ Å is shown in the left panel of Figure 4.2 as the solid line labeled $(1, 1)$ (this notation will be explained in the next section). Models characterized by different values of mean surface brightness are obtained by simple translations with respect to this reference model.

Note that in the idealized case of identical sources, an entire population of such sources is represented by a single point in diagrams such as Figure 4.2. Those lying on the indicated line in the diagram would be sufficient to ionize the IGM. Observations that fail to detect sources at some surface density–luminosity combination along the line would not rule out the model but instead could indicate that the sources lie at some other point along the line. The shaded area below the line indicates the area where sources would not produce enough ionizing photons for reionization.

The results for a different effective temperature are illustrated in the right panel of Figure 4.2, which is analogous to the left panel but refers to sources with $T_\star = 5 \times 10^4$ K.

4.2.5
Mean Metallicity at Reionization

It is interesting to compute the mean metallicity to which the universe is enriched by stellar populations with the minimum mean surface brightness required to ionize the IGM. Here, we define metallicity as $Z = \Omega_Z / \Omega_b$, where Ω_Z is the fractional comoving density in metals (in both galaxies and the IGM). Massive stars with effective temperatures in the range 3×10^4 to 10^5 K produce $\sim 10^{62}$ Lyman continuum photons per unit of stellar mass. Assuming a typical yield in met-

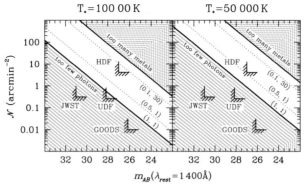

Fig. 4.2 Minimum surface brightness of reionization sources. The left panel is for Population III stars with an effective temperature of 10^5 K, while the right panel is for stars with an effective temperature of 5×10^4 K. In each panel we given the minimum areal number density of sources of a given AB magnitude at the wavelength corresponding to a 1400 Å rest frame. The upper shaded areas correspond to overproduction of metals in the IGM, while the lower shaded areas correspond to underproduction of ionizing photons. The models are parametrized by a couple of values (f, C_ϱ) where f is the escape fraction of ionizing radiation and C_ϱ is the gas clumping factor. The L-shaped markers indicate the combination of limiting magnitude and area for existing surveys with HST and for a possible future survey with JWST (Reproduced by permission of the AAS).

als of 0.2 [300], we find that the massive stars responsible for reionization enrich the universe to a minimum metallicity of $Z_{min} \sim 1.2 \times 10^{-4} Z_\odot$. This result does not change by more than a factor of 3 for a wide range of stellar properties. As an example, by considering only supermassive Population III stars, with masses between 100 and $1000 M_\odot$ with the yields as given by [113], and a power-law mass function with a slope between -1 and -2, we find that the mean mass-weighted yield is $\sim 0.2 \pm 0.05$ (neglecting the metals trapped in black hole remnants). Stars in this mass range will produce 7.6×10^{61} ph $M_\odot^{-1}$ [282], so that the minimum surface brightness model enriches the universe to a mean metallicity $Z_{min} \approx 1.7 \times 10^{-4} Z_\odot$. For a given IMF, any reduction of the output of ionizing photons can be regarded as an inefficiency in the energetics of reionization and will increase the fraction of hydrogen that needs to undergo nuclear processing and therefore will increase the mean metallicity. Thus, the maximum efficiency, minimum surface brightness model also corresponds to the minimum metallicity case.

Estimates of the mean density of heavy elements in the universe at the end of reionization ($z \simeq 6$) place additional constraints on the mean surface brightness of the stellar component of reionization sources. It is difficult to determine Ω_Z precisely since some metals will be in stars, some in the interstellar media of galaxies, and some in the IGM. But we can bracket the likely range of metal densities following an argument developed by Michael Fall. Songaila [247] has estimated a mean metal density in the IGM of $\Omega_{Z,IGM} \sim 3 \times 10^{-7}$, corresponding to $Z \sim 3 \times 10^{-4} Z_\odot$, at $z \simeq 5$ from the statistics of CIV and SiIV absorption-line systems. This provides a lower limit to the mean metal density shortly after the end of reionization. This limit is about twice the metallicity of the minimum surface brightness model Z_{min} (discussed above).

The present mean metallicity in galaxies is close to solar [132] and the present mean density of the stellar and interstellar components of galaxies is $\Omega_{gal} \approx 4 \times 10^{-3}$ [93]. Together, these imply $\Omega_{Z,gal} \sim 10^{-4}$ at $z = 0$. A variety of observations and models indicate that the fraction of mass in galaxies Ω_{gal} was smaller at $z \simeq 6$ than at $z = 0$ by a factor of 10 or more, and that the mean metallicity in galaxies was also smaller by a similar factor [209]. From these considerations we infer an upper limit on the mean metal density of $\Omega_Z \leq 10^{-5}$, corresponding to $Z_{max} = \Omega_Z/\Omega_b \leq 0.01 Z_\odot$ at $z \simeq 6$. This value is ~ 60 times higher than the metallicity of the minimum surface brightness model.

If metal-producing sources are surrounded by local HII regions, their luminosity at $\lambda = 1400$ Å can be enhanced by up to a factor of 3 due to the nebular continuum contribution. Including this factor, we use the limit to metal production as an upper limit to the mean surface brightness of reionization sources. This is obtained for $z_2 = 6$ and $\Delta z = 1$. Reionization at higher redshifts or occurring over a more extended time interval, produces the same (or higher) metallicity for fainter sources. Thus, the regions under the global metallicity constraint lines in Figure 4.2 define necessary but not sufficient conditions to guarantee that metals are not overproduced. All reionization models compatible with the global metallicity constraint must lie in the nonshaded region in Figure 4.2. Our results can be compared to those by Miralda-Escude and Rees [170] who found that galaxies producing a metallicity $Z = 10^{-2} Z_\odot$ would have a surface brightness of the order of $AB = 32$ mag arcsec^{-2}. Our minimum metallicity is smaller by a factor $\sim 1.7 \times 10^{-2}$ than the value considered by these authors. Converting $AB = 32$ mag arcsec^{-2} to mag arcmin^{-2}, correcting for our lower metallicity and dimming the surface brightness at rest frame 1400 Å by the factor ~ 4.5 that lets us convert from stars with $T_* = 5 \times 10^4$ K to stars with $T_* = 10^5$ K, we find values of surface brightness of $AB = 29$ mag arsec^{-2} that are broadly compatible with those in our Table 4.1.

4.3
Adding Realism to the Calculations

In the previous section we have implicitly assumed that all ionizing photons escape their sources and that the Universe is homogeneous. In this section we will abandon both of these assumptions.

4.3.1
Escape of Ionizing Photons

In the local Universe it is generally true that ionizing stars are located in HII regions, i.e. regions where an ionization–recombination equilibrium has been established. This fact implies that only a fraction of the ionizing UV radiation of these stars escapes to infinity. It is likely that even at high redshift some Lyman continuum photons are absorbed within the reionization sources themselves. Even in an environment with primordial composition, and thus free of dust, a lower escape fraction f_c can result if the individual sources are surrounded by an envelope of neutral hydrogen left by incomplete star formation in protogalaxies. In such clouds, recombinations would occur at much higher rates; these objects would produce local HII regions for at least part of their lifetimes and thus could display hydrogen and helium emission lines in their spectra [39, 109]. We have seen in Section 3.3.2 that a $300 M_\odot$ Population III star is able to completely ionize all gas in a halo with mass up to $M_s = 2.31 \times 10^6 M_\odot$ as given by (3.34). For $M \geq M_s$ the escape fraction is zero but for $M < M_s$ the escape fraction is given by $f_c = 1 - M/M_s$ and, e.g., for a $300 M_\odot$ star in a $10^6 M_\odot$ halo is $f_c \simeq 0.66$. For more massive halos we can compute the escape fraction including also the multiplicity $N_s(M)$ due to multiple stars forming in the same halo following (3.52). In this case we compute f_c as:

$$f_c = 1 - \frac{M}{M_s N_s(M)} \tag{4.14}$$

In Figure 4.3 we show the escape fraction predicted using (4.14) for stars with mass $M_\star = 150$ or $300 M_\odot$. It is clear that when stellar multiplicity is taken into account, high values of escape fraction are possible for both masses.

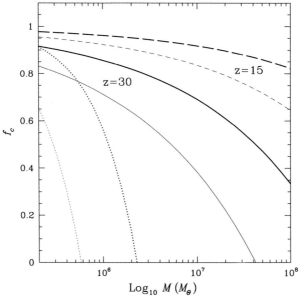

Fig. 4.3 The escape fraction of ionizing photons for Population III stars in mini-halos. The thick lines refer to Population III stars with a mass of $300 M_\odot$, while the thin lines are for $150 M_\odot$. The solid lines are for $z = 30$, while the dashed lines are for $z = 15$. The dotted lines are for individual Population III stars without multiplicity.

The case of galaxies is much more complex and we do not know the best value of the mean escape fraction. It only takes a column density of 1.7×10^{17} hydrogen atoms cm^{-2} to produce unit optical depth at the Lyman limit ($\lambda = 912$ Å). Moreover, some amounts of dust may well be present because of enrichment by previous generations of stars. Steidel et al. [258] derive from their observations, corrected for intervening IGM absorption, a reduction of the Lyman continuum with respect to the continuum at 1500 Å by ~50%. They estimate that dust absorption in the far-UV could be a factor of ~5. The total combined effect of gas and dust gives $f_c \approx 0.1$. Thus, these observations indirectly suggest that metal-free, dust-free galaxies might have $f_c \approx 0.5$. However, the properties of the reionization sources at $z > 6$, including the Lyman-continuum escape fraction, may be different from those of the Lyman-break galaxies at $z \simeq 3$. The analysis of a more recent sample [237] suggests a lower value $f_c \approx 0.02$, still implying $f_c \approx 0.1$ in the dust-free case. Leitherer and coworkers [136] argue for escape fractions below 15% in bright starburst galaxies in the local universe. However, these observational studies measure the escape fraction near 912 Å, while the hydrogen ionization cross section decreases rather rapidly with decreasing wavelength so that even with optical depth $\tau = 2-3$ at 912 Å one could have a Lyman-continuum escape fraction greater than 20%, especially for supermassive stars that are hotter than local O stars [167, 203]. There have also been a few attempts to estimate the escape fraction theoretically based on idealized assumptions on the structure of the sources. The derived values of the escape fraction span a relatively wide range roughly centered around $f_c \simeq 0.1$ [72, 73, 227]. In the absence of a definite value, we will consider two possible escape fractions: $f_c = 0.5$ and $f_c = 0.1$.

In principle, the emission line intensities one obtains for a system with $f_c > 0$ are not identical to those obtained for the case $f_c = 0$ reduced by $1 - f_c$ and their intensity will depend on the geometry of the source. However, it is easy to check by running the Cloudy software [82] that the difference in line strengths and in the continuum does not exceed 20% for $f_c < 0.99$ [220]. Only when the vast majority of the photons escapes does one obtain results that are significantly dependent on geometry.

An escape fraction $f_c \neq 1$ increases the required mean surface brightness relative to the minimum value. This in turn increases the required value of B by a factor f_c^{-1}. Moreover, if the star is embedded in neutral hydrogen, the fraction of Lyman-continuum radiation that does not escape generates a nebular continuum (mostly two-photon emission) in the rest-frame nonionizing UV, which dominates the observed continuum flux at $\lambda = 1400$ Å [203]. Combining the two effects, we find that the actual continuum flux F at $\lambda = 1400$ Å is higher than the value for the complete escape of ionizing photons by the factor

$$\frac{F(f_c)}{F(1)} \simeq \frac{3 - 2f_c}{f_c} \qquad (4.15)$$

where the coefficients 3 and 2 account for the fact that the nebular continuum at 1400 Å for Population III HII regions is roughly twice as high as the stellar continuum.

4.3.2
Clumpy IGM

In addition to inhomogeneities near and within the sources, there will also be density fluctuations on larger scales. Such inhomogeneities will increase the recombination rate and act as sinks of ionizing UV photons. This effect can be included by adopting values of the clumpiness factor C_ϱ in (4.11) different from unity. For a direct comparison with the homogeneous case, we have considered the case of a fixed value of clumpiness by adopting the value of $C_\varrho = 30$ [101, 156] and recomputed the mean surface brightness and the Thomson optical depth τ_T for the same models that we considered for $C_\varrho = 1$. The resulting values are also given in Table 4.1. Based on the simulations by [101], the value of $C_\varrho = 30$ appears to be a reasonable choice for $z \simeq 6-7$. At higher redshifts, smaller values are expected. Thus, it is likely that the effective value of C_v is bracketed by the two extreme values ($C_\varrho = 1$ and 30) we consider here. For this reason we also consider the intermediate value $C_\varrho = 5$. The variation of ionization fraction with redshift implied by these models is shown in Figure 4.4.

As a result of gravitational instability, more reionization sources will form and the IGM will become more clumpy for decreasing redshift. Thus, the increase in the ionization and recombination rates will at least partially offset each other. To assess the importance of these effects we have considered also a simple model where the run of clumpiness C_ϱ with redshift and that of the star-formation rate with redshift are described by analytical functions inspired by a numerical model by Gnedin and Ostriker [101]. We shall refer to it as the GO model. The star-formation rate has been rescaled so as to achieve reionization at $z = 6$. For this particular model, star formation begins at very early time and increases in rate as the redshift decreases. The clumping factor also increases with decreasing redshifts but reachs a plateau at higher redshift. We have adopted the following equations:

$$SFR(z) = \begin{matrix} 10^{-4+0.42 \cdot (19-z)} & \text{if } z \geq 9, \\ 10^{0.2} & \text{otherwise} \end{matrix} \quad (4.16)$$

$$C_\varrho(z) = \begin{matrix} 10^{0.00273 \cdot (24-z)^2} & \text{if } z \geq 8, \\ 10^{1.47712+0.1204 \cdot (6.5-z)} & \text{if } z \leq 6.5, \\ 10^{0.69897+0.51877 \cdot (8-z)} & \text{otherwise} \end{matrix} \quad (4.17)$$

This model is listed in Table 4.1 as $C_\varrho = $ GO and is characterized by a low value of B similar to the models with short reionization or small clumping factor. The variations of clumpiness and star-formation rate are shown in Figure 4.5 and the run of ionized fraction as a function of redshift is shown as the dotted line in Figure 4.4. It is remarkable how this model is similar to the model with constant star formation and $\Delta z = 4$ (the short dashed line in the figure). The effective clumping factor C_ϱ for GO model is $C_\varrho \simeq 1.81$, i.e. with this value of C_ϱ and $\Delta z = 4$ one obtains the same B value as the GO model.

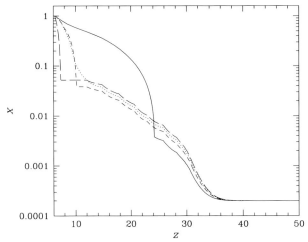

Fig. 4.4 The ionization fraction x as a function of redshift for a number of models. The solid and dashed lines are for constant star-formation models with clumping factor $C_\varrho = 5$. The solid line is for $\Delta z = 18$, the short dashed for $\Delta z = 4$ and the long dashed for $\Delta z = 1$. The dotted line is for a variable C_ϱ and star-formation rate model inspired by a simulation of Gnedin and Ostriker [101] and described in the text. For each model we start out with the primordial residual ionization fraction and we add a contribution to Population III stars (see Section 4.5). The Gnedin–Ostriker model produces an ionization fraction variation with redshift very similar to the model with $\Delta z = 4$.

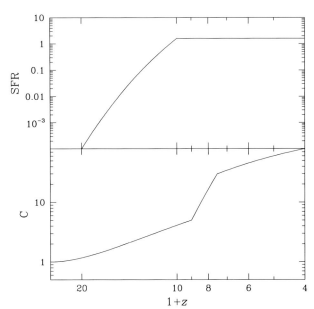

Fig. 4.5 The analytical fits to the star-formation rate (top panel) and run of clumpiness with redshift (bottom panel) for the simulation by Gnedin and Ostriker [101]

We find that the value of C_ϱ does not affect the optical depth τ_T significantly. In contrast, the surface brightness increases by a factor of 2–10 (i.e. 0.8–2.5 magnitudes) for the models with $C_\varrho = 30$, as shown in Table 4.1.

4.3.3
Two-Parameter Models

In the following, we denote models with the notation (f_c, C_ϱ) to indicate the adopted values of the Lyman-continuum escape fraction and of the clumpiness parameter. The (1, 1) label indicates the minimum surface brightness model as described in the previous section. The B factor of (4.5) is now a function of f_c and C_ϱ. Any model with $C_\varrho = 1$ located below the metallicity constraint in Figure 4.2 is allowed. In Figure 4.2, we plot as thin dotted lines the luminosity–surface density relations for identical reionization sources from models with $(f_c, C_\varrho) = (0.5, 1)$, and $(0.1, 30)$. The lines were computed for $z_2 = 6$ and $\Delta z = 1$ but are nearly identical for all $z_2 < 10$. The mean surface brightness of the reionization sources changes with Δz by less than a factor 4 for all $C_\varrho \leq 30$ (see Table 4.1). For the sake of completeness, we have explored a wider range of parameters than it is probably realistic. Note that both of the $(0.5, 1)$ and $(0.1, 30)$ models are allowed by the maximum metallicity constraint.

4.4
Luminosity Function of Ionizing Sources

So far, we have considered reionization sources with identical luminosity, but it is more likely that they are characterized by a broad luminosity function. On the basis of both theoretical considerations and experience at lower redshifts, it is reasonable to adopt a [233] differential luminosity function, $\Phi(L) \propto L^{-\alpha} \exp(-L/L_\star)$. Here, we adopt the parameters derived for Lyman-break galaxies at $z \simeq 3$–4, i.e. $\alpha = 1.6$ and $M_{\star,1400} = -21.2$ as our reference luminosity function [256, 305] and explore the effects of variations in both $M_{\star,1400}$ and α. The top left panel of Figure 4.6 was computed for the reference cumulative luminosity function in the case of $z_2 = 6$ and $\Delta z = 1$. The bottom solid line refers to the minimum surface brightness (1, 1) model. The area beneath this line would not reionize the IGM. The top solid line shows the metallicity constraints and the shaded area above this line shows the region where – with the adopted luminosity function – metals would be overproduced. The $(0.5, 1)$ and $(0.1, 30)$ models are shown as dotted lines. The remaining panels of Figure 4.6 explore the effects of varying $M_{\star,1400}$. The top right panel is for $M_{\star,1400} = -23.2$, i.e. two magnitudes of luminosity evolution compared to $z = 3$ Lyman-break galaxies. The bottom left panel is for $M_{\star,1400} = -17.5$ roughly corresponding to a star cluster with $10^6 M_\odot$ in massive stars forming over a period of ~ 1 Myr. Finally, the bottom right panel is for $M_{\star,1400} = -13.5$, which represents 7.7 magnitudes of luminosity evolution compared to the $z = 3$ Lyman-break galaxies and corresponds to the case of small dwarf galaxies or protoglobular clusters. The nonshaded area in

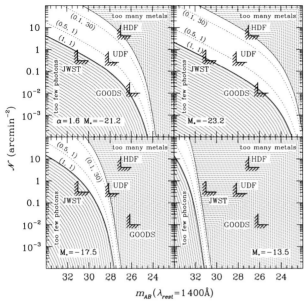

Fig. 4.6 Predictions on the luminosity function of ionization sources exploring changes in $M_\star$. For each panel we show the cumulative areal number density of sources brighter than a given AB magnitude measured at a wavelength corresponding to rest frame 1400 Å. The luminosity function slope is taken to be $\alpha = 1.6$ for all plots. In each panel the upper shaded area corresponds to overproduction of metals in the IGM, while the lower shaded area corresponds on underproduction of ionizing photons. The L-shaped markers show the combination of depth and area for various HST survey and for a hypothetical JWST deep survey. Excluding a portion of the nonshaded area rules out that particular luminosity function. It is clear that if $M_\star$ is small (e.g. lower right panel) than the HST surveys are not sensitive enough to detect the sources of reionization. Models are parametrized by the (f, C_ϱ) combination of escape fraction f and gas clumping factor C_ϱ (Reproduced by permission of the AAS).

the figures indicates the allowed region for each luminosity function. The curves would shift by less than 0.1 magnitudes for any value of $z_2 < 10$.

In contrast to those in Figure 4.2, the lines in Figure 4.6 represent luminosity functions that need to be populated everywhere in order to produce the required ionizing flux. Thus, a survey ruling out sources at one point along a curve would rule out that particular combination of (f_c, C_ϱ) model and luminosity function.

In Figure 4.7, we explore the effects of changing the slope of the luminosity function. The top left panel is identical to that of Figure 4.6 and is repeated here for convenience. The top right panel is for $\alpha = 1.1$, similar to the slope of the $z = 0$ luminosity function. The bottom left panel is for $\alpha = 1.9$. The bottom right panel of Figure 4.7 shows the forbidden region (double shading) and the preferred region (no shading) for a range of models. In this last panel, we have considered models with $M_{\star,1400}$ ranging from -21.2 to -16.2 and luminosity function slopes ranging from $\alpha = 1.1$ to 1.9. The lower shaded area represents the region of underproduction of ionizing photons common to all these luminosity functions. Similarly, the upper shaded area corresponds to the common region of overproduction of met-

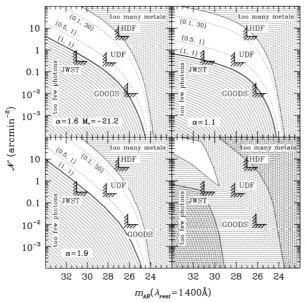

Fig. 4.7 Predictions on the luminosity function of ionization sources exploring the effect of changing the LF slope. The figure is the analog of Figure 4.6 and uses the same symbols and conventions. The value of $M_\star$ is fixed to -21.2 for all panels. The bottom panel illustrates which parts of this parameter space are the most promising to explore. The nonshaded area is the area common to all luminosity functions spanning a range in slopes $\alpha = 1.1$ to 1.9 and a range in $M_\star = -23.2$ to -17.5. A survey able to place constraints on this area would be the most predictive and a JWST deep surey appears to be capable of doing so. The UDF and GOODS are in the singly shaded area that has discriminating power for some luminosity functions but not for others (Reproduced by permission of the AAS).

als. The nonshaded area is allowed for all the models within the adopted range of parameters. Thus, this area should be favored in searches for the reionization sources.

The analogs of Figures 4.6 and 4.7 for models other than those with $z_2 = 6$ and $\Delta z = 1$ need in general to be recomputed by properly renormalizing the luminosity function. However, these curves can be obtained approximately by translating vertically the curves for the (1, 1) model with $z = 6$ and $\Delta z = 1$ by an amount $\Delta \log \mathcal{N} = -0.4 \Delta \mu_{AB}$, where $\Delta \mu_{AB}$ is the difference between the values of mean surface brightness of the models considered. This procedure is only approximate because it does not properly average the value of $M_{\star,1400}$ over the redshift interval Δz being considered.

4.4.1
Detecting Lyman α from Ionizing Sources

The fraction $1 - f_c$ of the ionizing continuum that does not escape generates local HII regions around the reionization sources, making them potentially visible in the H recombination lines and, more faintly, in the HeI lines. Prior to full reionization, when the universe consisted of sources embedded in their own HII regions, Lyα photons were able to escape if these HII regions were big enough that their outer boundaries were redshifted out of resonance [109, 157]. In the absence of dust, the fraction of Lyman α photons that escape from the HII region is determined solely by its size and hence by the production rate of ionizing photons of its central source $\dot{N}_c$ [140]. In the more realistic case, where some dust is present, an even a smaller fraction of Lyman α photons would escape [30, 52, 202]. Neglecting absorption by the neutral medium, the Lyman α line intensity can be computed as follows. Each recombination within the HII region has a ~2/3 probability of producing one decay leading to Lyman α emission (see Section 2.4). Under conditions of very low metallicity ($Z < 10^{-2} Z_\odot$), the Lyman α production is increased by ~50% due to collisional excitation [203]. The number density of ionizing photons absorbed within the local HII region is obtained from the required number density of photons needed to reionize the universe, n_c as $(f_c^{-1} - 1)n_c$, where f_c is again the escape fraction of Lyman-continuum photons. Following Miralda-Escudé and Rees [168] – and in analogy to (4.5) – we find that the rate of production of Lyman α photons per unit solid angle is:

$$\frac{d\dot{N}_\alpha}{d\Omega} = \frac{c}{4\pi}(f_c^{-1} - 1)n_c \tag{4.18}$$

The Lyman α surface brightness is the product of $d\dot{N}_\alpha/d\Omega$ and the mean observer-frame photon energy $h\bar{\nu}_\alpha$. For $\Delta z = 1$ and $z_2 = 6$, this surface brightness is $(f_c^{-1} - 1) \times 1.276 \times 10^{-16}$ erg cm^{-2} s^{-1} arcmin^{-2}. This value is a strong function of C_ϱ, decreases with increasing z_2, and depends only slightly on the duration of reionization. We now define f_α as the escape fraction of Lyman α photons. The average surface brightness in Lyman α, J_α, of the reionization sources corresponding to the (f_c, C_ϱ) model is then expressed as

$$J_\alpha = \frac{c}{4\pi} B(f_c^{-1} - 1)f_\alpha h\bar{\nu}_\alpha n_H \tag{4.19}$$

The effects of Lyman α scattering by intergalactic neutral hydrogen has been studied by a number of authors [157, 169]. Here, we will follow the approach by Santos [230].

The absorption cross section σ_N is given by [206, 230]:

$$\sigma_N(\nu) = \frac{3\lambda_\alpha^2 A_{21}^2}{8\pi} \frac{(\nu/\nu_\alpha)^4}{4\pi^2(\nu - \nu_\alpha)^2 + (A_{21}^2/4)(\nu/\nu_\alpha)^6} \tag{4.20}$$

where $\nu = c/\lambda$ and $\nu_\alpha = c/\lambda_\alpha$ with λ_α the wavelength of Lyman α.

Neutral-hydrogen atoms in the intergalactic medium will have nonzero thermal velocities and will redshift in and out of resonance. This leads to a thermal broadening of the line profile that can be modeled as a convolution of the absorption cross section with the velocity distribution function, namely:

$$\sigma_V(\nu) = \int_{-\infty}^{\infty} F(v)\sigma_N\left(\nu - \nu_\alpha \frac{v}{c}\right) dv \tag{4.21}$$

where the velocity distribution $F(v)$ is a Maxwellian distribution function:

$$F(v) = \left(\frac{m_H}{2\pi kT}\right)^{1/2} \exp\left(-\frac{m_H v^2}{2kT}\right) \tag{4.22}$$

where T is the temperature of the intergactic medium. The resulting cross section for Lyman α absorption is shown in Figure 4.8.

The optical depth for Lyman α photons is given by:

$$\tau(\nu) = \int (1 - x(r)) n_{HI}(r) \sigma_V(\nu(1 + v/c)) \, dr \tag{4.23}$$

where x is the ionized fraction and n_{HI} the neutral-hydrogen density, and with the integration carried out along the line of sight. Local ionization will affect $x(r)$. For a source with an intrinsic, normalized, line profile $\phi^{in}(\nu)$ the observed line profile is:

$$\phi^{obs}(\nu) = \phi^{obs}(\nu) \exp[-\tau(\nu)] \tag{4.24}$$

Integration of $\phi^{obs}(\nu)$ gives the total transmitted Lyman α so that the transmitted Lyman α fraction is:

$$f_\alpha = \int \phi^{obs}(\nu) \, d\nu \tag{4.25}$$

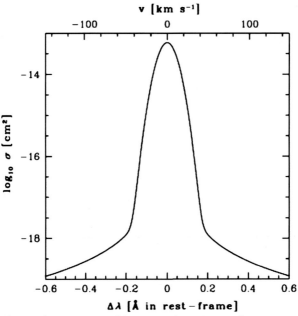

Fig. 4.8 The cross section for Lyman α absorption as a function of wavelength (bottom) and velocity (top) difference from resonance in the frame of the absorbing atom. The central Doppler core has a width determined by the gas temperature. The profile outside the core is known as the 'damping wings' (Reproduced by permission of the AAS).

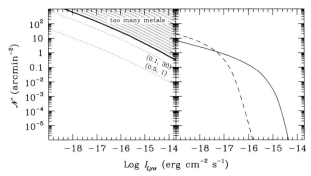

Fig. 4.9 Model predictions for Lyman α sources. The right panel shows the locus of identical reionization sources for two choices of parameters (f, C_ϱ) (dotted lines). The upper shaded area represents overproduction of metals. The area corresponding to underproduction of ionizing photons is now shown here because Lyman α luminosity can be made arbitrarily low by increasing the escape fraction of ionizing photons and still reionizing hydrogen. The right panel shows two luminosity functions for Lyman α sources capable of reionizing hydrogen for the case (0.5, 1). The solid line corresponds to $\alpha = 1.6$ and $M_\star = -21.3$, while the dashed line corresponds to $\alpha = 1.1$ and $M_\star = -17.5$. For both models the attenuation of Lyman α photons has been computed assuming an intrinsic width of 150 km s^{-1}.

We have computed [157, 169] f_α, assuming a line width of 150 km s^{-1}. In the left panel of Figure 4.9, we show the locus of identical Lyman α sources corresponding to $(f_c, C_\varrho) = (0.5, 1)$ and $(0.1, 30)$ and for $z_2 = 6$ and $\Delta z = 1$. We also show the metallicity constraint as the upper shaded area. In the case of Lyman α sources, there is no lower shaded area, indicating insufficient production of ionizing photons, as sources with f_c close to unity have arbitrarily low Lyman α flux but are still capable of ionizing the IGM. The right panel of Figure 4.9 shows two luminosity functions for the Lyman α sources. The solid line represents the same luminosity function as the $z = 3$ Lyman-break galaxies converted to Lyman α luminosity for the (0.5, 1) model. The dashed line represents a luminosity function with $\alpha = -1.1$ and $M_{\star,1400} = -17.5$.

In addition to the case of an intrinsic width of 150 km s^{-1}, we have also considered cases with 75 to 300 km s^{-1}. As expected, the wider the line, the more Lyman α radiation escapes. For example, for sources at the faint end of Figure 4.9, the fraction of escaping Lyman α photons is $\sim$ 30% higher for $v = 200$ km s^{-1}. We will revisit in more detail the escape of Lyman α photons in Chapter 5.

4.5
Reionization by Population III Stars

The optical thickness constraint from WMAP [252] suggests that a sizeable fraction of hydrogen was ionized by redshift 10. It is reasonable to compute what is the expected contribution to reionization of Population III stars. We can estimate this by assuming the Population III cumulative rate in the presence of a Lyman–

1 Cosmic Reionization

Tab. 4.2 Difference in Thompson optical depth τ_T when reionization by Population III stars is also taken into account. Reionization is completed at $z = 6.2$ in all cases. The table gives the redshift interval over which reionization takes place Δz (col 1), the clumpiness parameter C_ϱ (col 2), and the optical depth τ_T without (col 3) or with (col 4) pre-reionization by Population III stars.

Δz	C_ϱ	$\tau_T^{no\ PopIII}$	$\tau_T^{with PopIII}$
1.0	1.0	0.045	0.052
1.0	5.0	0.045	0.052
1.0	30.0	0.046	0.050
4.0	1.0	0.057	0.065
4.0	5.0	0.060	0.065
4.0	30.0	0.067	0.069
18.0	1.0	0.090	0.096
18.0	5.0	0.101	0.104
18.0	30.0	0.108	0.109
18.0	GO	0.059	0.066

Werner background derived in Section 3.2.2. From the cumulative number of Population III stars and their lifetime, we can derive their total production of ionizing photons per unit volume. When compared to the total number of hydrogen atoms in the same volume this gives us the ionization fraction that could be attributable to Population III stars if recombinations were negligible and the escape fraction for Population III stars was unity. This gives us an ionization fraction ≈ 0.06 at $z = 10$. In our calculation we will adopt for the Population III component the same number of ionizing photons per hydrogen atom B adopted for the model being considered. Table 4.2 gives for the same models as Table 4.1, the values of the Thompson optical depth τ_T with or without the contribution to reionization of Population III stars with density given by Figure 3.4. Comparing these entries it is clear that reionization by Population III adds only 0.008 or less to the value of τ_T. These calculations have been made assuming an escape fraction of unity. Any smaller value will reduce the impact of Population III stars even further.

4.6
How Is the Intergalactic Medium Enriched?

When considering the mean metallicity of the Universe at reionization, we have been referring to the total amount of metals produced. This is in general very different from the metallicity actually measured in the intergalactic medium (IGM). If Population III stars are formed in halos of sufficiently low mass they can enrich the IGM by SN-driven winds [158, 177] (see also Section 3.3.3). When a halo undergoes a SN-driven outflow, the ejection of metals can be very effective. However, it is not clear how effective this process is when averaged over all halos.

Note that if the escape fractions of UV radiation and metals are comparable, we should expect the mean metallicity of the IGM at reionization to be similar to the mean metallicity of the Universe, $Z \simeq 10^{-3} Z_\odot$ (see Sections 4.1 and 4.2.5). This is broadly consistent with the metallicity of damped Lyman α systems that is higher than $10^{-3} Z_\odot$ for $z < 4$ [217].

The IGM probably continues to be enriched also following reionization as observations of the properties of Lyman-break galaxies at $z \sim 3$ seem to indicate [4] and we should not expect to be able to constrain Ω_Z at high redshift uniquely on the basis of reionization.

4.7
Reheating of the Intergalactic Medium

Ionizing sources have a spectral-energy distribution extending smoothly beyond the Lyman continuum and, as a consequence, they will have a mean energy of their ionizing photons exceeding by some finite amount the hydrogen ionization energy. Values of the excess energy for blackbodies within a range of temperatures were given in Table 3.1. The excess energy available after a hydrogen atom is ionized is going to be distributed between the proton and the electron, with the electron gaining the vast majority of it when considering the conservation of energy and linear momentum in the recoil. In the presence of collisions the two energy distributions of electrons and protons will thermalize and the two species will evolve toward the same temperature with the final outcome being a reheating of the intergalactic medium (IGM). We have discussed similar processes in Section 3.3.1 for a halo that is denser than the IGM. Several processes will drive the evolution of the IGM temperature including adiabatic expansion, and cooling by helium, however, at very high z the dominant mechanism is inverse Compton cooling on the cosmic microwave background (CMB) photons [61]. This is because, in contrast to the thermal equilibrium within an HII region discussed in Section 3.3.1, the much lower density makes all mechanisms dependent on the square or cube of the density much less efficient.

The cooling function per electron can be written as:

$$\Lambda = 5.5 \times 10^{-36}(1+z)^4[T - 2.7(1+z)] \text{ erg s}^{-1} \qquad (4.26)$$

where the term $T - 2.7(1 + z)$ accounts for the difference in temperature between the electrons and the CMB, while the term in $(1 + z)^4$ accounts for the increase with redshift of the energy density of CMB photons. The temperature associated to any of the excess energies of Table 3.1 is far in excess of $2.7(1 + z)$ K so that for all practical purposes we can neglect the CMB temperature.

The cooling function allows us to derive a cooling timescale:

$$\tau_{Compton} \equiv \frac{3 k_B T_e}{2 \Lambda} \simeq 4.07 \times 10^{13} \text{ s} \left(\frac{1+z}{31}\right)^{-4} \qquad (4.27)$$

We can obtain an indication of how effective this cooling mechanism is by computing the ratio of the inverse Compton cooling rate timescale divided by the local

Hubble time (see, e.g., (2.38)), namely:

$$\frac{\tau_{Compton}}{t_H} = 0.0166 \left(\frac{1+z}{31}\right)^{-5/2} \quad (4.28)$$

Equation (4.28) implies that at high-redshift inverse Compton cooling is very fast and cools the electrons on timescales much shorter than the local Hubble time. This remains true for $z \gg 5$ as at $z \simeq 5$ the Compton cooling time and the Hubble time are identical. In practice, other physical processes will become important [61, 274].

Inverse Compton cooling could in principle introduce distortions into the CMB spectrum. To first order one could argue that these distortions are small, given that electrons are only a fraction $\sim 10^{-10}$ of the CMB photons. However, each of these electrons has an energy (see Table 3.1) much larger than the mean CMB photon energy $\sim 2.3 \times 10^{-4}(1+z)$ eV. Thus, in principle, one could generate small but detectable distortions [71].

A similar reheating of the IGM is expected to take place in correspondence to the reionization of helium. Under the assumption that the reionization of helium takes place at low redshift by measuring the IGM temperature before and after the reionization of helium one could verify the model on the basis of the reheating associated to helium reionization and constrain the epoch of reionization of hydrogen [272]. This will be revisited in Chapter 8.

4.8
Keeping the Intergalactic Medium Ionized

So far we have been concerned with reionizing hydrogen. However, a different criterion can be derived by requiring that hydrogen remains ionized, i.e. that new ionizations balance recombinations. It is possible to estimate the ionizing flux needed to keep the Universe reionized by requiring that all hydrogen is reionized over a recombination timescale [156]. Clearly this quantity depends critically on the assumed value of the clumpiness parameter C_ϱ. For large values of C_ϱ the ionizing flux needed to keep the Universe ionized may coincide with that required to reionize it in the first place.

The ionization rate is obtained from the requirement of ionizing all hydrogen during one recombination timescale. In terms of the comoving production rate density of ionizing photons thus becomes:

$$\dot{n}_c = \frac{n_H}{t_{rec}} \quad (4.29)$$

where n_H is the total hydrogen (comoving) density and the recombination timescale t_{rec} is given by:

$$t_{rec} \simeq \left(n_H(1+z)^3 \alpha_B C_\varrho\right)^{-1} \quad (4.30)$$

where we have taken into account the dependence of t_{rec} on the physical, rather than comoving, hydrogen number density. Combing (4.29) and (4.30) with $n_H \simeq 1.8 \times 10^{-7}$ cm^{-3} and making use of the value of α_B from (2.11) for $T = 2 \times 10^4$ K, we find:

$$t_{rec} \simeq 3.3 \times 10^{15} \text{ s} \left(\frac{1+z}{7}\right)^{-3} \tag{4.31}$$

Using the value from (4.31) into (4.29) now gives:

$$\dot{n}_c \simeq 1.87 \times 10^{-20} \text{ s}^{-1} \text{ cm}^{-3} \left(\frac{1+z}{7}\right)^3 C_{30} \tag{4.32}$$

where $C_{30} \equiv C_\varrho/30$, or in comoving Mpc^3:

$$\dot{n}_c \simeq 5.48 \times 10^{53} \text{ s}^{-1} \text{ Mpc}^{-3} \left(\frac{1+z}{7}\right)^3 C_{30} \tag{4.33}$$

In order to derive the ionizing photon density relevant for this case we need to multiply the production rate of (4.32) by a timescale that is in this case given by the mean lifetime of ionizing photons. This is given by the attenuation length Δl divided by the speed of light c.

Let us compute the attenuation length. Following Madau et al. [156] we can write the attenuation optical depth between redshift z and z_0 as:

$$\tau_{eff} = \frac{4}{3}\sqrt{\pi S_0} N_0 \left(\frac{\nu}{\nu_c}\right)^{-1/2} (1+z_0)\left[(1+z)^{3/2} - (1+z_0)^{3/2}\right] \tag{4.34}$$

where $S_0 = 6.3 \times 10^{-18}$ cm^{-2} is the hydrogen photoionization cross section at the Lyman edge ν_c and N_0 is a parameter fixed to reproduce the observation of hydrogen-absorption systems.

Equation (4.34) can be expanded near a given redshift z_0 to find at the Lyman edge:

$$\tau_{eff} \simeq 2\sqrt{\pi S_0} N_0 (1+z_0)^2 \Delta z \simeq 0.356(1+z_0)^2 \Delta z \tag{4.35}$$

Requiring now $\tau_{eff} = 1$ in (4.35) gives us:

$$\Delta z \simeq 2.81(1+z)^{-2} \tag{4.36}$$

We can convert this value of Δz to a proper distance (computed as the comoving line of sight distance divided by $1+z$) to find:

$$\Delta l \simeq 3.68 \text{ Mpc} \left(\frac{1+z}{7}\right)^{-9/2} \tag{4.37}$$

We can now write the ionizing intensity per unit frequency (or ionizing surface brightness) seen by an observer at zero redshift as:

$$J_c \simeq \frac{1}{4\pi} \dot{n}_c h \Delta l \tag{4.38}$$

where we have neglected factors of order unity that account for the specific spectral shape [156]. The numerical value that we derive is:

$$J_c \simeq 3.25 \times 10^{-25} \text{ erg s}^{-1} \text{ cm}^{-2} \text{ Hz}^{-1} \text{ sr}^{-1} \left(\frac{1+z}{7}\right)^{3/2} C_{30}^{-1} \quad (4.39)$$

This value is somewhat smaller than what is typically found in the literature because of a combination of factors such as that in ΛCDM cosmology Δl is longer and that at $T = 2 \times 10^4$ K, as adopted here, the recombinations are less effective than at $T = 10^4$ K typically adopted in the literature. The flux of (4.39) corresponds roughly to a surface brightness of $AB = 30.3$ arcmin^{-2}, which would need to be corrected from the rest-frame wavelength of 912 Å to the observed wavelength for a suitable spectral-energy distribution of the sources. For blackbody spectra the correction term to a rest-frame wavelength of 1400 Å ranges from 0.6 mag for $T_* = 1.1 \times 10^5$ K to –0.6 for $T_* = 3 \times 10^4$ K. Clearly, this simple treatment ignores, in the case of hard ionizing spectra, the contribution from secondary ionizations and also Auger electrons that would become important as the IGM metallicity increases [110, 240].

4.9
Hints for Further Study

- For a given density, recombinations are more frequent when structures form and increase the clumping factor of gas. At the same time the density of gas decreases with redshift so that expansion and structure formation partly compensate each other. Is there a redshift when reionization was the easiest?
- Assuming that we want to reionize hydrogen but not helium II and that the bulk of reionization is done by Population II stars, what is the maximum contribution of AGNs? Similarly, what is the maximum contribution of Population III stars?
- Assuming an escape fraction of continuum photons $f_c = 0.1$ and a Lyman α escape fraction $f_\alpha = 0.5$, what is the surface brightness of the diffuse Lyman α emission if reionization was completed by Population II stars at $z = 6$ after a $\Delta z \simeq 3$? how would that change if reionization was done by Population III stars or by solar metallicity stars?

Observational Techniques and their Results

5
Studying the Epoch of Reionization of Hydrogen

5.1
Overview

This chapter is devoted to a discussion of the major observational techniques to investigate the reionization history of hydrogen. The large absorption cross section of neutral hydrogen to Lyman α photons is the basis of several techniques. The most direct method is based on the fact that neutral hydrogen distributed uniformly along the line of sight would absorb Lyman α radiation resonantly at its rest-frame wavelength, which corresponds to a range of wavelengths in the spectrum of a background source. This absorption trough shortwards of Lyman α in the spectrum of high-redshift QSOs is known as the Gunn–Peterson [103] trough and has been a classical diagonostics tool for reionization since it was originally proposed in 1965. The method and its limitation are discussed in Section 5.2. Instead of using bright QSOs as background sources one could probe the presence of diffuse neutral hydrogen by studying the evolution of the luminosity function of Lyman α sources. In this case, the diagnostic for each source is local and one is not sensitive to neutral hydrogen over a redshift interval but at the precise redshift of each source. This technique is presented in Section 5.3.

Neutral hydrogen at the low temperature that it will have before reionization (tens of K) is mostly found in the ground state. Even at these temperatures the hyperfine spin-flip level responsible for 21-cm radiation can be excited and, depending on the specific physical conditions, one can expect 21-cm emission from neutral hydrogen. Alternatively, one could observe 21-cm absorption when the level is radiatively excited by the cosmic microwave background radiation. Both effects can in principle be used to study the history of reionization as a function of redshift and are discussed in Section 5.4.

Instead of studying reionization by tracing neutral hydrogen one can also study it by looking at its products. Indeed, reionization produces free electrons and they can be used as diagnostics by constraining the scattering of the cosmic microwave radiation. This is discussed in Section 5.5.

From First Light to Reionization. Massimo Stiavelli
Copyright © 2009 WILEY-VCH Verlag GmbH & Co. KGaA, Weinheim
ISBN: 978-3-527-40705-7

5.2
Gunn–Peterson Troughs in Redshift 6 QSOs

QSOs are the ideal system to probe intervening absorption in the intergalactic medium (IGM) as their spectra are characterized by a relatively featureless continuum with emission lines but no significant absorption lines (clearly this is not true for broad absorption line QSOs but these represent a minority of QSOs) and they are very bright and found up to high redshift. Any given QSO acts as a background source enabling us to study all Lyman α absorption systems up to the redshift of the QSO along that particular line of sight. Clearly, multiple QSOs need to be observed to obtain a result averaged over multiple lines of sight and thus more representative.

5.2.1
A Simple Gunn–Peterson Test

In the idealized case reionization of hydrogen would be seen in the spectra of QSOs as complete absorption longwards of Lyman α. In fact, for a system embedded in

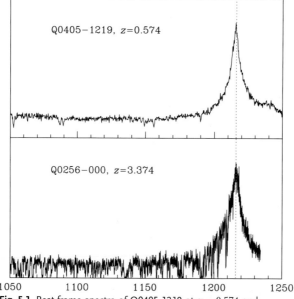

Fig. 5.1 Rest-frame spectra of Q0405-1219 at $z = 0.574$ and Q0256-000 at $z = 3.374$. Both spectra are from the QSO spectral library made available by Jill Bechtold [14, 15]. The continua and broad emission lines of the two QSOs are similar but it is clear how the higher-redshift QSO has a much larger number of Lyman α absorption systems.

a neutral medium we have [80, 103]:

$$\tau_{GP} = 1.8 \times 10^5 h^{-1} \Omega_M^{-1/2} \left(\frac{\Omega_b h^2}{0.02}\right) \left(\frac{1+z}{7}\right)^{3/2} \frac{n_{HI}}{n_H} \quad (5.1)$$

which is extremely high for a neutral medium and indeed a very low neutral hydrogen fraction is sufficient to obtain significant optical depth (see Figure 4.1).

The presence of an increasingly large number of discrete neutral hydrogen clouds along the line of sight is a complicating factor difficult to control. This is illustrated in Figure 5.1 where we compare the spectra of a low-redshift QSO at $z < 1$ with that of a QSO at $z \approx 3$. While the low-z QSO only displays a handful of Lyman α absorbers, the high-redshift system shows a rich complement of absorption systems. The spectral appearance of these systems has become known as the Lyman α forest. For even higher redshift it is natural to expect even more absorption systems with the intrinsic shape of the continuum becoming increasingly difficult to measure.

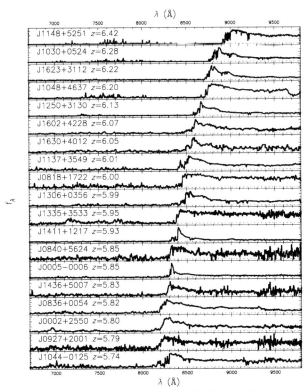

Fig. 5.2 Rest-frame spectra of QSOs at redshift around 6 discovered by the Sloan Digital Sky Survey [80]. It is clear that continuum absorption increases with redshift but it is also apparent how hard it is to understand whether the absorption by multiple systems that we know is present is also combined with that from a diffuse neutral component (Reproduced by permission of the AAS).

As shown in Figure 5.2, by the time one reaches redshift $z \simeq 6$ the absorption systems are so numerous that the continuum shortwards of Lyman α is almost completely absorbed, masking the possible presence of a diffuse neutral medium. This complicates the implementation of the simple Gunn–Peterson test and requires more complex diagnostics.

5.2.2
The Gunn–Peterson Trough

Absorption by a neutral medium is expected to be so significant that the wings of the absorption would extend beyond Lyman α, generating a characteristic Gunn–Peterson trough and a damping wing. While the clear detection of the trough and of the damping wings would be an unmistakable signature of reionization, not detecting them leaves open multiple interpretations. In particular, even before reionization, a QSO can be bright enough to ionize the medium in its immediate neighborhood, thus removing the gas that, in the ideal case, would be responsible for the trough.

The simple formula of (5.1) is valid at the resonance but does not take into account that significant absorption is possible also off-resonance, thanks to the presence of the damping wings. We can compute the Gunn–Peterson optical thickness including the effect of the wings as [157]:

$$\tau_{GP}(\lambda) = \int_{z_{reion}}^{z_{em}} \frac{dl}{dz} n_{HI}(z)(1+z)^3 \sigma_N\left(\nu = \frac{c(1+z)}{\lambda}\right) dz \qquad (5.2)$$

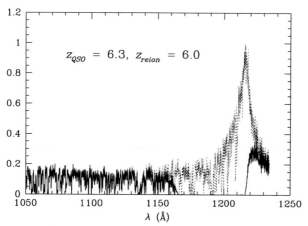

Fig. 5.3 Illustration of the effect of a neutral medium on a $z = 6.3$ QSO. The spectrum is actually that of Q0256-000 redshifted to $z = 6.3$ [14] and it has therefore far fewer Lyman α absorbers than a real QSO at $z \approx 6$. The solid line shows the spectrum as observed if reionization occurs at $z = 6$. The thin dotted line gives for reference the unattenuated spectrum.

where σ_N is the absorption cross section given in (4.20) and n_{HI} is the comoving neutral-hydrogen number density. The transmission at a given wavelength is given by $\exp(-\tau)$. In Figure 5.3, we simulate the observation of a QSO at $z = 6.3$ embedded in a neutral medium reionized at $z = 6$. The effect of the Gunn–Peterson trough is quite evident. Comparison with the unattenuated spectrum also shows that the damping wing significantly reduces the strength of the Lyman α line.

By varying the function $n_{HI}(z)$ it is possible to use (5.2) to evaluate the proximity effect caused by local reionization by the QSO. This is illustrated in Figure 5.4 where we show the transmission for an object at $z = 6.3$ with reionization at $= 6$ and compare it to the case when the object is surrounded by an ionized bubble of 2 or 4 Mpc in radius. Due to the very extended damping wings of the cross section σ_N, in a neutral medium there is significant absorption even to the red of Lyman α. However, a local ionized bubble of a few Mpc in size – as shown in the figure – can enable significant Lyman α transmission.

When determining the value of τ_{GP} from real data, one needs to exclude the wavelengths that are affected by proximity effects. Similarly, one has to stop the analysis at the wavelengths affected by Lyman β and OVI emission, i.e. in practice 1040 Å in the QSO rest frame. Clearly, because of the large optical depth, one needs very high S/N spectra to measure a transmission or place a meaningful upper limit to it. One also needs to take into account the cumulative effect of discrete Lyman α absorbers. A fit to QSOs at $z < 5.5$ gives [80]:

$$\tau^{eff} \simeq 0.85 \times \left(\frac{1+z}{5}\right)^{4.3} \tag{5.3}$$

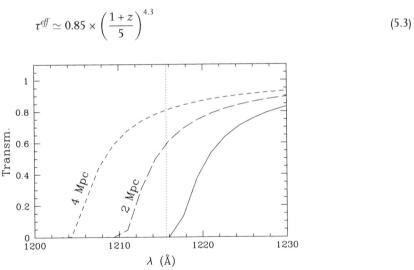

Fig. 5.4 Transmission in the Gunn–Peterson trough versus rest-frame wavelength for a source at $z = 6.3$ embedded in a neutral medium reionizing completely at $z = 6.0$ (solid line). The long dashed line illustrates the effect of an ionized bubble 2 Mpc in size centered on the object, while the short dashed line is for a 4-Mpc bubble. The vertical thin dotted line indicates the position of Lyman α.

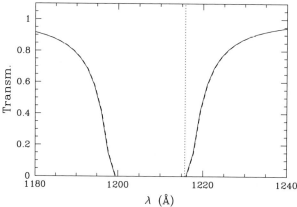

Fig. 5.5 Transmission in the Gunn–Peterson trough versus rest-frame wavelength for a source at $z = 6.3$ for a model where reionization begins at $z = 24$ and is completed at $z = 6$. This curve is indistinguishable from that derived for the case of a rapid reionization beginning at $z = 7$ and still ending at $z = 6$ (dashed line). The vertical thin dotted line indicates the position of Lyman α.

Measuring τ_{GP} in excess of this value is an indication of the presence of diffuse neutral hydrogen.

The large cross section of neutral hydrogen for Lyman α absorption makes the observed transmission also relatively independent of the reionization history. In Figure 5.5 we show the transmission for a QSO at $z = 6.3$ for a model where reionization is completed at $z = 6$ but begins at $z = 24$. In this model the residual neutral-hydrogen fraction at $z = 6.3$ is only $\sim 1.6 \times 10^{-2}$ but this value is still sufficient to produce a quite opaque Gunn–Peterson trough. This curve would be indistinguishable in this plot from one for reionization beginning at $z = 7$ and still ending at $z = 6$.

5.2.3
Lyman Series Lines

Other lines of the Lyman series have smaller cross section than the Lyman α line. The optical depth in Lyman β is ≈ 6 times smaller than that in Lyman α, while the optical depth in Lyman γ is ≈ 18 times smaller than that of Lyman α. As a consequence, the Lyman β and Lyman γ forest lines are less saturated than the Lyman α forest lines and represent potentially more powerful diagnostics. Unfortunately, their wavelengths are affected by lower-redshift Lyman α absorbers and this makes the analysis more complex. The combined analysis on Lyman α, β, and γ has been carried out on the SDSS QSOs [80] and is shown in Figure 5.6, where it is clear that there is strong evidence for accelerated evolution with redshift for $z > 5.5$. This interpretation has been challenged on the basis of its dependence on a particular probability distribution for the absorbers optical depth. Changing from a density

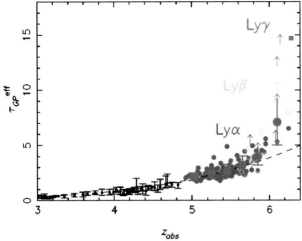

Fig. 5.6 Measured values and lower limits to the Gunn–Peterson optical thickness in Lyman α, β and γ as a function of redshift for SDSS QSOs [80]. The extrapolation from lower redshift accounting for the cumulative effect of individual hydrogen clouds is given by the dashed line. There is evidence at $z \approx 6$ for an optical thickness in excess of the extrapolation (Reproduced by permission of the AAS).

distribution [171] to, e.g., a log-normal distribution would change the extrapolation and weaken the evidence for reionization at $z \simeq 6$ [17].

5.2.4
Metal Lines

A way to bypass the difficulties implicit in the large optical depths of hydrogen lines is to rely on metal lines. Especially at low metallicities these will be unsaturated and therefore will represent powerful tracers of the neutral fraction [183]. One of the most promising lines in this respect is OI at 1302 Å, which is just to the red of Lyman α. Observing the OI forest would probe the topology of reionization and metal pollution of the IGM.

5.2.5
HII Region Size Test

It is possible to estimate the size of the local HII region ionized by the QSO. The size of this HII region is expected to depend on the fractional density of neutral hydrogen in the surrounding medium as more ionizing photons will be needed to expand the local HII region into a neutral medium compared to a partially ionized medium.

This method suffers from a number of uncertainties having to do with the QSO lifetime, uncertainty in the ionizing flux and the contribution of nearby galaxies

110 | *5 Studying the Epoch of Reionization of Hydrogen*

to the ionizing flux. The latter effect could be particularly serious if QSOs are surrounded by an excess of galaxies [266].

5.2.6
Dark Gaps

Near the redshift or reionization one expects extended wavelength regions of high optical depth caused by the 'skin' of ionized bubbles that have not yet percolated. These regions would appear as dark gaps in the spectrum. Using dark gaps as a tracer of reionization requires detailed comparison of their statistical properties to models. Galaxies close to the line of sight would alter the dark-gap statistics and also need to be modeled.

5.2.7
An Assessment of the Indication from QSOs Spectra

Combining all the diagnostics described in the previous subsections provides a consistent picture suggesting the presence of neutral hydrogen along the line of sight

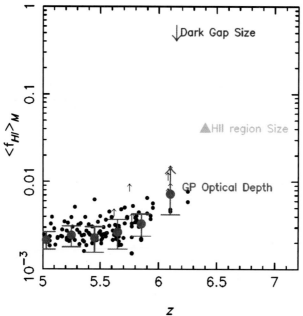

Fig. 5.7 Constraints on the neutral-hydrogen fraction as a function of redshift. The black points represent individual GP optical depth measurements from QSOs, while the large red points are averages. The green triangle represents constraints from the HII region size, while the downward blue arrow represents the constraint from the dark-gap size (Reproduced by permission of the AAS).

of QSOs at $z > 6$ [80]. This is illustrated in Figure 5.7, where we show the constraints on the neutral-hydrogen fraction from the various diagnostics. The neutral fraction has to be larger than 0.01 to be consistent with the observed Gunn–Peterson troughs.

5.3
Lyman α Sources as Diagnostics of Reionization

Radiation at the wavelength of Lyman α is absorbed resonantly by neutral hydrogen so that in principle one could derive a redshift for reionization by looking for rapid changes in the luminosity function of Lyman α sources. Naively, one would expect the sources in place before reionization to be significantly absorbed. In practice, one expect some Lyman α emission to escape whenever the intrinsic line width is large enough, as the photons in the red wing have a lower probability of being absorbed. Another effect is that a Lyman α source is very likely also a source of ionizing radiation and this radiation could create a bubble devoid of neutral hydrogen in the vicinity of the source. If large enough, this bubble could ensure that the line emission escapes unabsorbed. In the following we will consider all these effects.

5.3.1
Effect of a Finite Lyman α Line Width

The core width of the effective absorption cross section is determined by the temperature of the intergalactic medium (see Figure 4.8 and (4.21)). However, the core width plays a very minor role unless the neutral-hydrogen fraction (or, equivalently, the neutral-hydrogen column density) is low. For the typical cases of interest the temperature of the IGM setting the width of the absorption core does not play a significant role.

If the intrinsic Lyman α width is large, some of the photons will be emitted out of resonance and will be able to make it to the observer without being absorbed. It is clear by looking at Figure 5.3 that even for the broad Lyman α of a QSO, absorption is significant and thus we should not expect a major effect for narrower lines. This is illustrated in Figure 5.8 where we have assumed the intrinsic line profile to be a Gaussian with dispersion σ and we show the Lyman α escape fraction f_α as a function of σ. Clearly the effect of the line width is modest. One would need line widths in the 1000 km s^{-1} range to have a sizeable effect.

5.3.2
Intrinsic Properties of Lyman α Emitters

Each generation of stars produces metals that enrich the Universe. The mean metallicity of the Universe decreases with increasing redshift and it is likely that this also applies to the mean metallicity of the Lyman α emitters. However, we have seen in Figure 3.10 that the intrinsic equivalent width (hereafter EW) of the

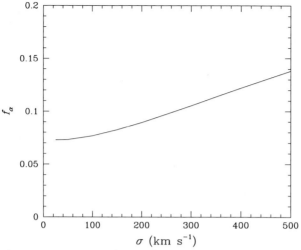

Fig. 5.8 Lyman α escape fraction f_α plotted versus intrisic dispersion σ in km s^{-1} of the line. For definiteness we have assumed that the emitter is at $z = 6.5$ with a residual neural-hydrogen fraction of 0.03 and that the line profile is a Gaussian with the given σ. It is clear that the escape-fraction dependence of σ is only modest for the values of interest for a galaxy.

Lyman α line increases with decreasing metallicity. Thus, at early times, an intrinsically stronger line could partly mask the effect of increased absorption adding a complication to the use of Lyman α emitters as diagnostics for the reionization epoch. We can model this effect by assuming a scenario broadly compatible with both the SDSS QSO data and the WMAP optical depth result and where reionization is completed at $z = 6$ but begins at $z = 24$. For simplicity, we adopt a neutral-hydrogen fraction that varies linearly within this interval and assume that the Universe is enriched in metals proportionally to its ionizing photons output so that metallicity is also linear over this redshift range. In Figure 5.9 we compare the ionization fraction x as a function of redshift for this model – with and without a contribution from Population III stars – to the model with $\Delta z = 18$ from Chapter 4. The two start reionization at the same redshift and have similar properties. The neutral fraction at $z \simeq 6.1$ is about 0.01, in agreement with the constraint in Figure 5.7.

We start from a primordial metallicity at $z = 24$ and consider two end values of metallicity at reionization, namely $0.1 Z_\odot$ or $0.01 Z_\odot$. For the strength of Lyman α as a function of metallicity we assume that the galaxy has the same mean metallicity as the Universe and adopt a fixed initial mass function (IMF) with stars from 1 to $100 M_\odot$ to link metallicity to Lyman α intensity using the models by Schaerer [232]. Our results are given in Figure 5.10 and show a relatively modest dependence on redshift, with a variation within a factor of two. The real effect could be larger if, for instance, the metallicity of the Lyman α emitters was lower than the mean of

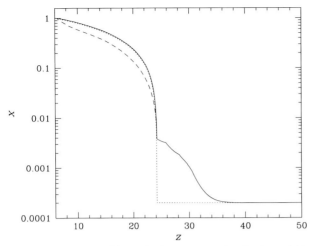

Fig. 5.9 Comparison of the ionization x as a function of redshift for a simple model with ionization increasing linearly with redshift from $z = 24$ to $z = 6$ with a model with constant rate of ionization over the same redshift interval. The solid line gives x for the simple model including the contribution of Population III stars, while the dotted line is without Population III contribution. The dashed line is for the model with a constant rate of ionization and the same redshift interval ($\Delta z = 18$).

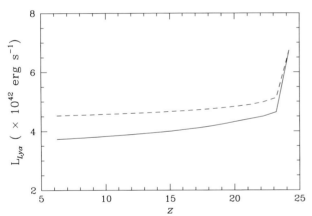

Fig. 5.10 Lyman α luminosity as a function of redshift for a galaxy with a star-formation rate of $1 M_\odot$ yr^{-1}. The metallicity is varied as a function of redshift by assuming it to be equal to the mean metallicity of the Universe at that redshift. The variation in luminosity for a given star-formation rate is less than a factor of two.

the Universe and the IMF depended on metallicity. The former would be plausible, e.g., if the Lyman α emitters we see were galaxies undergoing their first major burst of star formation, producing the bulk of their stars and of their metals. The latter is most likely true for primordial metallicities where one might well have a truncated IMF extending up to much larger masses, but we do not know whether it applies also to small but nonprimordial metallicities. Another effect that we neglect is that of dust. As metallicity increases, dust can be present and the presence of a dense dusty neutral region within the source itself might affect more dramatically Lyman α than the ionizing continuum because of the resonant nature of the line that will undergo many scattering events accumulating a longer path within the dusty neutral medium.

5.3.3
Effect of a Local Ionized Bubble

A bright Lyman α emitter is also likely to have ionized the surrounding volume of space. In order to have an impact on the escape of Lyman α photons the size of the ionized volume must be large enough that the Hubble flow over that distance is sufficient to bring Lyman α photons out of resonance. We have seen an indication of this effect in Figure 5.4 where we show how the Gunn–Peterson trough is modified by the presence of local bubbles. In Figure 5.11 we show the Lyman α escape fraction as a function of the ionized bubble radius. This is computed by convolving the transmission as a function of wavelength with the Lyman α line shape (still assumed to be a Gaussian). We find that large ionized bubbles are required to have high values of f_α but even relatively small Mpc-sized bubbles can enable small leakage at the 10% level. Obviously, a given bubble size is easier to achieve if the medium is already partly ionized. For the sake of this calculation, partial ionization outside the large ionized bubbles could be due to many, small, nonoverlapping ionized bubbles or due to a contribution of X-rays from, e.g., AGNs that would partly reionize hydrogen at large distances from the source.

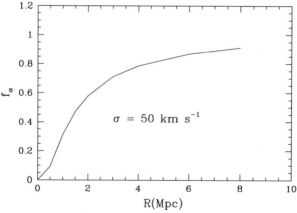

Fig. 5.11 Lyman α escape fraction f_α as a function of the radius in Mpc of a local fully ionized bubble. We have assumed an intrinsic Lyman α width of 50 km s^{-1}. Even for such a narrow line the presence of a relatively small ionized bubble can enable some Lyman α leakage.

5.3.4
A Realistic Lyman α Escape Model

In the previous subsections we have considered a number of effects having an impact on the observed luminosity of Lyman α sources. Let us consider now a combined model where all these effects come into play. We will consider the slow reionization scenario beginning at $z = 24$ and ending at $z = 6$. Galaxies will have intrinsically stronger Lyman α for decreasing metallicity, following Figure 5.10, and will be able to surround themselves with a small reionized bubble produced by a fraction f_c of their ionizing photons. The remaining $1 - f_c$ ionizing photons are involved in recombinations leading to the Lyman α luminosity in about two thirds of the cases. The reionized bubbles will enable increased Lyman α escape following a mechanism similar to that shown in Figure 5.11. In Figure 5.12, we show the Lyman α attenuation in this model for a galaxy with a star-formation rate (SFR) of 20 $M_\odot$ yr^{-1}, an age of 100 Myr, an escape fraction of ionizing photons $f_c = 0.1$, and a velocity dispersion of 200 km s^{-1} and for a galaxy with the same age but a SFR of only 2 $M_\odot$ yr^{-1}, an escape fraction $f_c = 0.35$ and a velocity dispersion of 100 km s^{-1}. The specific attenuation is largely the product of the SFR by the galaxy age, so a galaxy producing 2 $M_\odot$ yr^{-1} for 100 Myr would have an attenuation similar to one with SFR 20 $M_\odot$ yr^{-1} case but a lifetime of only 10 Myr. The figure suggests that Lyman α escape fractions above 30–40% should be observable even before

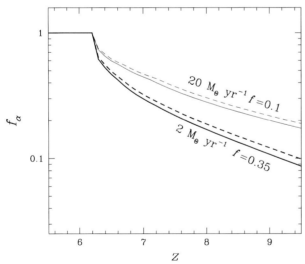

Fig. 5.12 Lyman α escape fraction f_α as a function of redshift for the slow reionization model described in the text. All sets of curves have been obtained assuming a galaxy age of 100 Myr. The thick curves refer to the case of a galaxy with a star-formation rate of 2 $M_\odot$ yr^{-1} and an escape fraction $f_c = 0.35$, while the thin curves are for 20 $M_\odot$ yr^{-1} and an escape fraction $f_c = 0.1$. The solid lines are for a final metallicity at reionization of $0.1 Z_\odot$, while the dashed lines have a final metallicity of reionization of $0.01 Z_\odot$. The predicted Lyman α transmission at $z = 6.5$ ranges from 38 to 55%.

reionization. The major effect of the presence of a diffuse neutral medium will be more modest at the bright end of the luminosity function but dramatic at the faint end where a significant flattening of the luminosity function should be observable. Unfortunately, it is difficult to convert this qualitative prediction into a firm quantitative one because the size of the local ionized bubble, which drives the escape fraction of Lyman α, depends not only on the Lyman α luminosity but also on the age of the source and on the contribution of neighbors that might not be Lyman α emitters.

Let us consider the type of luminosities we would predict at $z = 6.5$, which is a redshift that can be more easily explored from the ground thanks to an atmospheric window. The intrinsic Lyman α luminosity of a source with SFR $2 M_\odot \text{ yr}^{-1}$ would be $\sim 3 \times 10^{42} \text{ erg s}^{-1}$, for the type of metallicities we expect at $z \approx 6$ (0.01–0.1$Z_\odot$) and considering $f_c = 0.35$. After correcting for attenuation of about 60% one finds at $z = 6.5$ a luminosity $\sim 1.2 \times 10^{42} \text{ erg s}^{-1}$. This compares well to the luminosity of the typical bright Lyman α emitters that are observed at $\sim 4 \times 10^{42} \text{ erg s}^{-1}$ but intrinsically fainter sources with similar lifetimes would be attenuated more significantly and a large attenuation would lead to a strong evolution of the luminosity function from $z = 6.5$ to $z = 5.7$. Thus, the small evolution of the bright end in the observed luminosity function from $z = 6.5$ to $z = 5.7$ [160] is roughly compatible with our scheme. Our model relies on a number of assumptions, namely:

- The lifetime of the Lyman α emitting galaxies 100 Myr. A shorter lifetime would decrease the size of the local ionized bubbles increasing the observed attenuation.
- We have assumed line widths of 100 and 200 km s^{-1}, but smaller line widths would not significantly alter our results. Larger line widths would reduce the attenuation.
- For the escape fraction of ionizing photons we have assumed $f_c = 0.1 - 0.35$. In particular, the latter value may be on the high side (see discussion in Section 4.3.1). A lower escape fraction would reduce the size of the ionized bubbles and increase the attenuation.
- The evolution of intrinsic properties of the Lyman α emitters from $z = 5.7$ to $z = 6.5$ is modest. Faster evolution would go in the direction of reducing the attenuation as a rapid build up in metallicity and possibly dust at $z \simeq 5.7$ would increase the effective Lyman α strength at $z = 6.5$ compensating for a larger attenuation.
- In our model the diffuse medium is 97% ionized at $z = 6.5$. A lower ionized fraction would increase the attenuation. A medium 99% ionized would increase the Lyman α transmission from 38–55% to 52–67%. A medium only 90% ionized would decrease it to 21–37%.
- We have ignored the effect of clustering of sources around each bright Lyman α emitter. Clustering would increase the size of the fully ionized bubbles.

The escape fraction f_c is one of the least-known parameters and has potentially major implications. An escape fraction $f_c \ll 0.1$ within the framework of our model would imply that the slow evolution with redshift of the Lyman α luminosity func-

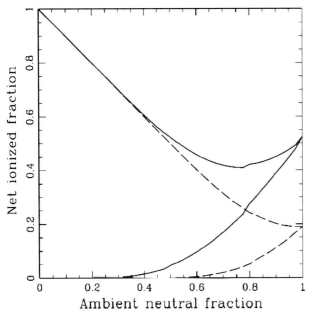

Fig. 5.13 Total ionization fraction (upper curves) and volume fraction in fully ionized bubbles (lower curves) required to enable sufficient escape of Lyman α photons as a function of the ambient neutral fraction. The solid lines refer to the case of zero line width, while the dashed lines are for a line width of 360 km s^{-1} [162] (Reproduced by permission of the AAS).

tion would be difficult to reconcile with the results from the Gunn–Peterson trough of SDSS QSOs.

Our conclusions are based on a specific model for the evolution of Lyman α sources and their metallicity as well as of the ionization fraction of the Universe. Evidence in favor of significant ionization of hydrogen can be obtained in a relatively model-independent way following the method of Malhotra and Rhoads [162]. The idea is to constrain the fraction of ionized-hydrogen volume that must be present in order to allow Lyman α sources located at random positions not to be excessively attenuated. By requiring an attenuation no larger than a factor of 2, this gives an ionized volume fraction in bubbles of about 20–50%. This fraction could decrease if the ambient medium is partly ionized, as shown in Figure 5.13, therefore this analysis is broadly compatible with our model.

5.3.5
Perspectives on Studying Reionization with Lyman α Sources

The discussion so far shows that Lyman α sources are potentially a good tracer of reionization but their use as a precise diagnostic is hampered by model uncertainties. There are, however a number of options to improve their discriminating power. We have already seen that the escape fraction depends critically on the source luminosity in any model where the local bubbles are the dominant factor in determining the escape of Lyman α photons. We show in Figure 5.14 the change in luminosity function that this implies. We adopt the luminosity function at $z \simeq 5.7$ with faint end slope $\alpha = 1.5$ and use the model described in the previous subsection to predict how it would appear at $z \simeq 6.5$ due to the finite transmission of Lyman α. The predicted change is a reduction in normalization and a flattening of the faint end slope to 1. This flattening could be turned into a diagnostic if we could refine the measurement of the faint end slope at $z \simeq 5.7$ and measure it accurately at $z \simeq 6.5$. This would require deeper observations than presently available but should be or become feasible in the near future. A change in slope by $\Delta\alpha = 0.5$ over the relatively short time from $z = 5.7$ to $z = 6.5$ would be difficult to explain as intrinsic evolution and would point to the presence of a diffuse partly neutral medium. On the other hand, the nondetection of this change would be a strong indication that reionization was completed at higher redshift or that local bubbles do not play an important role.

A conceptually simple way to bypass model-dependent impacts on Lyman α transmission would be to study – as a function of redshift – the ratio of Lyman α to a Balmer line such as Hα or Hβ. The Balmer lines are to first order proportional to the Lyman α intensity, e.g., Hα tracks Lyman α to better than 1% for $Z < 0.05 Z_\odot$, but are not affected by the presence of a neutral medium. This is not a precise correction because lines can have collisional contributions but it is still useful. Unfortunately, observing Balmer lines for high-redshift galaxies in the reionization era

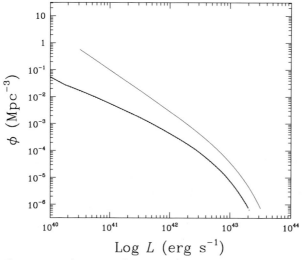

Fig. 5.14 How the escape of Lyman α variation in the model discussed in the text affects the luminosity function. The thick line represents the predicted luminosity function at $z = 6.5$, while the thin line represents the original luminosity function [160]. The faint end slope changes from 1.5 to about 1.

is beyond the capabilities of present telescopes and will require the next generation of facilities currently under development.

Another important diagnostic would be to study Lyman α sources at higher redshift [199]. Detecting individual sources is important as they provide a proof of existence for these galaxies. However, it is important to place the emphasis on samples as it is difficult to draw any conclusion from individual Lyman α sources especially when considering geometric effects within a source [52] in addition to all the other uncertainties discussed so far.

5.3.6
Faint Lyman α Halos

There are at least three different physical mechanisms able to produce extended Lyman α emission. The first is related to the Lyman α attenuation that we have discussed in this section. So far we have considered the absorption of Lyman α photons by neutral hydrogen only as a process reducing the source luminosity. However, under low-dust conditions, the absorbed photons are simply reradiated in a random direction with the resulting effect of generating a diffuse, low surface brightness Lyman α halo.

The second mechanism is related to the cooling of a dark halo. We have seen in Chapter 3 that sufficiently massive halos cool by atomic hydrogen. The main energy-loss mechanism for these halos is Lyman α emission and this should appear as a diffuse low surface brightness halo. The third mechanism is Lyman α fluorescence produced by ionization fronts around QSOs at the edge of reionization. In the following, we will briefly discuss these mechanisms.

Scattering halo around a Lyman α source In principle, detection of this halo for a high-redshift source would enable us to estimate the fraction of neutral hydrogen present near the object. One would start from the observed position of the Lyman α peak. An asymmetric peak would be an indication of Lyman α attenuation and one would expect the diffuse emission to be to the blue of the observed peak. Because of the extent of the damping wings the size of the halo would be of the order of several Mpc. Indeed, Figure 5.4 shows about 60% transmission for an object surrounded by a 2-Mpc ionized bubble and 80% for a 4-Mpc bubble. Thus, about 30% of the line intensity is scattered at distances between 2 and 4 Mpc. At $z = 6$ in our cosmology 1 arcsec subtends roughly 6 kpc so that the diffuse halo would have a radius of ~8 arcmin. Assuming that half the flux of a 10^{43} erg s^{-1} source is scattered into a diffuse halo the resulting mean surface brightness would be ~1.5×10^{-22} erg s^{-1} cm^{-2} arcsec^{-2}, which would be very challenging to detect. Despite this difficulty, detecting the diffuse halo would be very important as it would allow us to estimate the Lyman α transmission and therefore constrain the neutral hydrogen fraction at the source.

Lyman α cooling halo We can express the virial temperature of a halo given in (2.57) as:

$$T_{vir} \simeq 17548\,K \left(\frac{M}{10^8 M_\odot}\right)^{2/3} \left(\frac{1+z}{10}\right) \quad (5.4)$$

Thus, the total kinetic energy of baryons in such a halo is:

$$K \simeq 6 \times 10^{52}\,\text{erg} \left(\frac{M}{10^8 M_\odot}\right)^{5/3} \left(\frac{1+z}{10}\right) \quad (5.5)$$

We can estimate the hydrogen cooling time τ_α as $kT_{vir}/(n_H \Lambda(T))$. Adopting $\Lambda(T) \simeq 10^{-22}$ erg s^{-1} cm^3 and (3.9) for the hydrogen number density n_H, we find:

$$\tau_\alpha \simeq 7.4 \times 10^{11}\,\text{s} \left(\frac{M}{10^8 M_\odot}\right)^{2/3} \left(\frac{1+z}{10}\right)^{-2} \quad (5.6)$$

which is extremely fast. The resulting cooling luminosity is K/τ_α. If we assume that 2/3 of the energy is in the Lyman α line, the line luminosity would be:

$$L_\alpha \simeq 5.4 \times 10^{40}\,\text{erg s}^{-1} \left(\frac{M}{10^8 M_\odot}\right) \left(\frac{1+z}{10}\right)^3 \quad (5.7)$$

Considering now the half-mass radius given in (3.37) we can estimate the surface brightness for such a cooling halo. Let us start at $z = 10$ where we find a surface brightness of $\sim 1.5 \times 10^{-19}$ erg s^{-1} cm^{-2} arcsec^{-2} and a small radius of ~ 0.28 arcsec. Such a halo should be detectable with, e.g., the James Webb Space Telescope (see Chapter 9).

The same $10^8 M_\odot$ halo at, e.g., $z = 2.4$ would have a surface brightness of $\sim 6.5 \times 10^{-20}$ erg s^{-1} cm^{-2} arcsec^{-2} and a radius of 0.43 arcsec. The surface brightness depends on $M^{1/3}$ so that even much more massive halos would remain difficult to detect with present instrumentation.

It is intriguing that some Lyman α 'blobs' have been detected [257] but they are much larger (100 kpc) and much brighter (10^{-15} erg s^{-1} cm^{-2} of integrated line flux) than the structure described here and it is at this stage still unclear what their origin is.

Lyman α Fluorescence A quasar ionization front expanding in the neutral medium will produce Lyman α emission [45]. The production rates from radiative recombinations and collisional excitation are shown in Figure 5.15. For temperatures higher than 10^4 K collisional excitations may become the primary channel of Lyman α production as long as the medium has an intermediate ionization fraction. Indeed, the collisional excitation rate depends on the product $n_{HI} n_e$ of the neutral-hydrogen density and the electron density. This implies that this rate is suppressed for a completely ionized or completely neutral medium but is important in the case of partial ionization. The predicted surface brightness is about 10^{-22} erg s^{-1} cm^{-2} arcsec^{-2} at $z \simeq 6.5$ with a redshift dependence proportional to $(1+z)^{-2}$. Given the large sizes of QSO ionization bubbles the emission should be visible over several hundred comoving Mpc^2, which translates into several hundred arcmin^2. The detection of Lyman α fluorescence may have already happened at $z \sim 3$ [5].

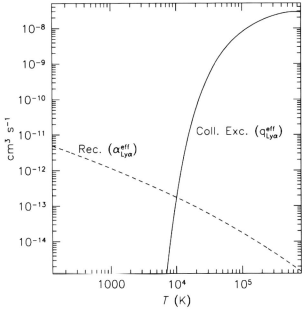

Fig. 5.15 Effective Lyman α production rates from recombinations (short dashed line) and collisional excitations (solid line) as a function of the gas temperature (Reproduced by permission of the AAS).

5.4 Neutral-Hydrogen Searches

In the nearby Universe, neutral hydrogen is studied primarily through the so-called 21-cm radiation. The same can in principle be used at high redshift [57, 118, 155, 238]. The hyperfine spin-flip energy-level splitting of the ground state of hydrogen corresponds to radiation of 21 cm wavelength, or 5.9×10^{-6} eV in energy, corresponding to a temperature of ~0.07 K. The state with antiparallel spins is a singlet and has lower energy than the state with parallel spins, which is a triplet. The ratio of the occupation number of the two states is then:

$$\frac{n_t}{n_s} = 3e^{-\frac{0.07K}{T_S}} \tag{5.8}$$

where n_t and n_s are the occupation numbers of the triplet and singlet states, respectively, and T_S is the so-called spin temperature. The spin temperature tends to be in thermal equilibrium with the CMB. However, in order to observe cosmological 21-cm lines one needs the spin temperature to differ from the CMB temperature. At high redshift, when the gas density is high, collisions couple the spin temperature to the gas temperature [218]. Integration of (2.12) gives us the gas temperature after the decreasing density and ionized fraction bring the gas out of thermal equibrium from the radiation for $z \lesssim 200$. We show the gas and CMB temperatures in

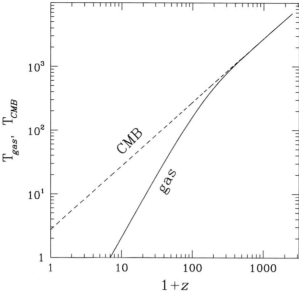

Fig. 5.16 Comparison of the gas temperature (solid line) and the CMB temperature (dashed line) as a function of redshift. Gas and CMB decouple thermally at $z \approx 200$ and after that redshift gas begins to cool and continues to do so in this calculation where the formation of galaxies is ignored.

Figure 5.16. Since the gas temperature and the spin temperatures are lower than the CMB temperature, at $z \lesssim 200$ the 21-cm line can be seen in absorption over the CMB. At lower redshifts the ionizing radiation from stars and galaxies heats up the gas, making it warmer than the CMB. The scattering of Lyman α photons now is able to couple the spin temperature to the gas temperature and the 21-cm line can appear in emission [84, 85, 301].

Neutral-hydrogen density variations will produce fluctuations in the 21-cm line brightness, which will be seen in emission or in absorption depending on redshift. This is potentially a very promising method to study primordial density fluctuations on small scales [142]. At lower redshift, 21 cm could be used directly to study the reionization history. Indeed, even partial reionization ensures $T_s \gg T_{CBM}$ [309], which should enable the observations of the 21-cm line in emission. The brightness temperature through a patch of the IGM with optical depth τ_{21} is:

$$T_b = T_{CMB} e^{-\tau_{21}} + T_S (1 - e^{-\tau_{21}}) \tag{5.9}$$

where the first term accounts for the absorption and the second for the emission. The flux difference between the patch and the CMB is given by the differential antenna temperature:

$$\delta T_b \simeq \frac{T_S - T_{CMB}}{1+z} \tau_{21} \tag{5.10}$$

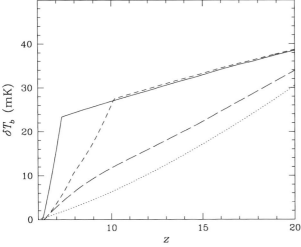

Fig. 5.17 Variation with redshift of the differential antenna temperature δT_b for constant ionization-rate models preionized by Population III stars. In all models reionization is completed at $z \simeq 6$. The solid line refers to fast reionization with $\Delta z = 1$, the short dashed line to $\Delta z = 3$ and the long dashed line to $\Delta z = 24$. The dotted line is a model with ionization fraction varying linearly with redshift between $z = 24$ and $z = 6$. A sharp step in δT_b is present only when reionization is fast but fast reionization is marginally at variance with the optical depth constraint by WMAP.

where τ_{21} depends on T_{CMB}/T_S, and is proportional to the neutral fraction $1-x$ and the overdensity of the patch δ and is given by [57, 155]:

$$\tau_{21} = \frac{3}{32\pi}\lambda_{10}^3 A_{10}\frac{0.07 K}{T_S}(1+\delta)(1-x)\frac{n_H}{H} \tag{5.11}$$

where $A_{10} = 2.85 \times 10^{-15}$ s^{-1} is the spontaneous decay rate of the transition and λ_{21} is the wavelength. The value of the differential antenna temperature is:

$$\delta T_b \simeq 16\,\text{mK}\left(1-\frac{T_{CMB}}{T_S}\right)(1+\delta)(1-x)\left(\frac{\Omega_b h^2}{0.02}\right)\left[\left(\frac{1+z}{10}\right)\left(\frac{0.3}{\Omega_m h^2}\right)\right]^{1/2} \tag{5.12}$$

It is worth noting that the antenna temperature given by (5.12) does not depend on T_S in the limit $T_S \gg T_{CMB}$.

A possible experiment would be to detect a global step in δT_b associated to reionization. In Figure 5.17 we show predictions for the differential antenna temperature as a function of redshift for the 3 constant ionization rate models introduced in Chapter 4 and the simple model of slow reionization introduced in this chapter. It is apparent that the slow reionization models that would be favored by the WMAP optical depth constraints do not display a sharp step in δT_b. However, being able to carry out this measurement would provide us with unique information on the reionization history of hydrogen.

Rather than just measuring the global variation of δT_b with redshift one might also measure the spatial variations of δT_b either directly or, more likely, through a correlation or an angular power spectrum analysis [91, 92].

5.4.1
Other Applications of High-z 21-cm Observations

In addition to probing directly the history of reionization, the 21-cm diagnostic could be applied to a number of other problems in high-z cosmology. We will highlight below a few of the most interesting applications.

Studying the fluctuations power spectrum The power spectrum of 21-cm anisotropies probes the power spectrum of primordial fluctuations at scales much smaller than feasible with CMB experiments [142]. In Figure 5.18 we show the power spectrum at $z = 55$ for 21-cm anisotropies for several different spectra. This measurement would allow us, for instance, to discriminate between a standard CDM scale-invariant spectrum and the case where dark matter is constituted by 1-keV neutrinos.

Studying the sources of reionization If reionization is dominated by low-mass galaxies, an interesting alternative to direct imaging of these faint objects could

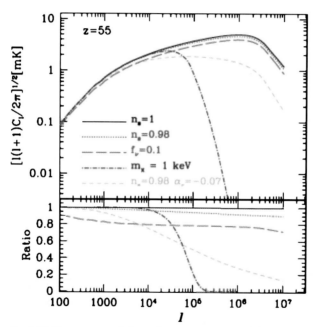

Fig. 5.18 The upper panel shows the power spectrum of 21-cm anisotropies at $z = 55$ for a ΛCDM scale-invariant model (solid black line), a model with $n = 0.98$ (dotted red line), a model with $n = 0.98$ and $\alpha_r = -0.07$ (short-dashed green line), a model with warm dark matter in the form of 1 keV neutrinos (dot-dashed blue line), and a model in which 10% of the matter density is in 3 species of massive neutrinos with mass 0.4 eV each (long-dashed purple line) [142]. The lower panel shows the radios between each of these spectra and the scale-invariant one (Reprinted with permission from [142]. Copyright (2004) by the American Physical Society).

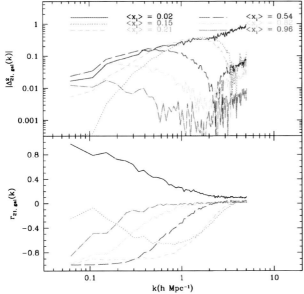

Fig. 5.19 The upper panel shows the absolute value of the 21-cm galaxy cross power spectrum for different redshifts and their corresponding ionization fractions as computed in a particular model of reionization [139]. The lower panel shows the corresponding cross-correlation coefficient (Reproduced by permission of the AAS).

be to study the correlation between fluctuations in the surface brightness of the Lyman α emission and the 21-cm signal from the IGM at the same redshift. This wold allow us to probe directly the connection between the sources of ionizing photons and the distribution of neutral hydrogen [303]. In Figure 5.19 we show the 21-cm galaxies cross-spectrum. After the early stages of reionization, which are more uncertain because the spin temperature T_S may not be much larger than the CMB temperature T_{CMB}, an anticorrelation is established between the 21-cm and, for instance, Lyman α emitting galaxies. The scale of this anticorrelation will evolve with the size of the ionized bubbles [139]. The details of the correlation will depend to some extent on the specific galaxy-selection technique used for the analysis. For instance, Lyman α selected galaxies will tend to be located at the center of large ionized bubbles so as to increase the Lyman α escape fraction. Their cross-correlation is then different from that of generic galaxies as shown in Figure 5.20.

Studying dark energy Fluctuations in the neutral-hydrogen distribution tracked by 21-cm radiation should reveal the acoustic peaks and therefore could be used as a way to measure baryon acoustic oscillations. Calculations show that this could be possible in the redshift range 1.5–6, which would be of interest to explore whether

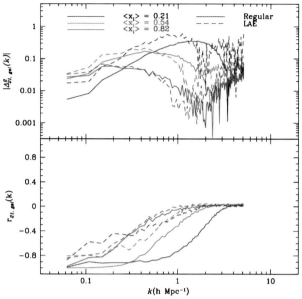

Fig. 5.20 The upper panel shows as a function of the wave number the absolute value of the 21-cm Lyman α galaxy cross power spectrum for different redshifts and ionization fractions (color coded). The lower panel shows the correlation coefficient. The difference between this figure and Figure 5.19 is in the selection of the galaxy sample, which is here limited to Lyman α emitters (Reproduced by permission of the AAS).

the observed acceleration of the Universe is a pure cosmological constant (which would be undetectable at high z) or a physical field [304].

5.5
Compton Optical Depth

The WMAP satellite detected statistically significant correlation between CMB temperature fluctuations and polarization. At angular scales $l > 20$ this correlation is in agreement with model prediction for CMB fluctuations. However, a large excess on scales $l < 10$ cannot be predicted uniquely on the basis of the power spectrum and was attributed to cosmological reionization [129, 130, 252]. In the absence of reionization CMB photons would arrive to us unaltered from the surface of last scattering. Observing polarization at large scales indicates that there was scattering of photons by electrons over a global scale. Evidence that a significant fraction of electrons has been scattered again after the 'last scattering surface' allows one to derive a value for τ_T and constrain the reionization epoch. The value obtained from the 5-year data is $\tau_T = 0.084 \pm 0.016$ [130], corresponding to $z_{reion} = 10.8 \pm 1.4$ in a simple step model of reionization.

As we have seen in this chapter, the optical depth τ_T is a powerful constraint to apply to our reionization models. We saw in Chapter 4 that reionization over $\Delta z = 1$ would produce $\tau_T = 0.045$ (see Table 4.1). This value is now excluded at $\sim 2.5\sigma$ by the 5-year WMAP measurements. $\Delta z = 3$ is now within 2σ of the observed τ_T, while slow reionization with $\Delta z = 18$ is within one sigma of the measured value.

5.6
Lyman α Signature in the Diffuse Near-IR Background

Somewhat similar to the idea of constraining the reionization history of hydrogen by studying the 21-cm line one could instead look for spectral features in the diffuse near-IR background generated by the Lyman α line [11, 238]. Strictly speaking one would be studying the history of recombinations rather than the history of reionization but the two can be interconnected. A few predicted spectra are shown in Figure 5.21 compared to sensitivity limits for various instruments on HST and JWST (NGST in the figure). It appears that this effect is going to be very difficult to detect even with JWST.

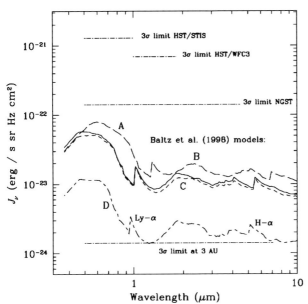

Fig. 5.21 Predicted spectra of the diffuse background for a number of reionization models (A–D) spanning different values of Ω_b and box size. The horizontal lines show observational limits by existing instruments [238]. NGST is the old name of JWST (with permission from AA Copyright (1999) ESO).

5.7
Hints for Further Study

- Adopting $z = 6$ as the redshift when reionization is completed and the simple model with constant ionization rate, what is the range of acceptable value for the start of reionization allowed by the 5-yr WMAP result? If one adopted for the ionization rate an exponential function of time with an e-folding time of 100 Myr – roughly compatible with the change in luminosity density at redshift $z \gtrsim 5$ – how would this change the acceptable range for the start of reionization? What if reionization is completed at $z = 7$?
- Adopting the luminosity function of Lyman-break galaxies at $z \simeq 6$ derived by Bouwens et al. [33], how does the radius of the fully ionized sphere change for an M_* galaxy if one considers the contribution from lower-mass galaxies associated to it? One can assume $f_c = 0.1$ and the spectral energy distribution of a 100-Myr old continuous star-formation object to estimate the ionizing-photon output of galaxies.
- Assuming that the luminosity function of Lyman-break galaxies $z \simeq 6$ derived by Bouwens et al. [33] describe also galaxies at $z \geq 6$ once the normalization is scaled down by a factor of 3, what is the total surface brightness in the J band of this population? What is the expected fluctuation assuming an angular correlation function slope of 0.5 and a correlation distance of 4 Mpc [135, 212]?

6
The First Galaxies and Quasars

6.1
Overview

Once we have determined that reionization took place at redshifts greater than 6, it is of great interest to try and identify the galaxies responsible for bringing it about. This is a difficult task and one necessarily dependent on assumptions on the nature of these sources. The primary difficulty is that we are by necessity forced to observe galaxies in their nonionizing continuum. Therefore, their contribution to the total ionizing flux can only be obtained as a result of an extrapolation. Such extrapolations are particularly uncertain. In fact, we cannot be sure of the spectral energy distribution of the galaxies we see at very high redshift and we do not know the value of the escape fraction of their ionizing photons. To these essentially physical limitations, we need to add another of a more practical nature, namely that, as we have shown in Chapter 4, galaxies maximally efficient in reionizing hydrogen would be the hardest to observe in the nonionizing continuum.

Despite these difficulties, several groups have attempted to identify galaxies that might be responsible for or might contribute to the reionization of hydrogen. Three main techniques for identifying galaxies at high redshift are commonly used: the Lyman-break technique, the Lyman α excess technique, the Balmer-jump (also known as 4000 Å break) search technique. In the following sections we will review each of these techniques and we will then discuss some results obtained with their application. We will also review the observational properties of the first stars, the implication of observed samples of galaxies and QSOs on the reionization of hydrogen and methods to constrain high-z galaxy populations based on the analysis of the unresolved background.

6.2
The Lyman-Break Technique

The Lyman-break technique is based on the identification of a discontinuity in the rest-frame ultraviolet continuum flux of a galaxy at high redshift generated by hydrogen absorption in the galaxy itself and by discrete neutral-hydrogen clouds

From First Light to Reionization. Massimo Stiavelli
Copyright © 2009 WILEY-VCH Verlag GmbH & Co. KGaA, Weinheim
ISBN: 978-3-527-40705-7

along the line of sight. This discontinuity is large enough that it can be identified through the use of a set of broadband filters.

In practice, the Lyman-break technique consists in identifying galaxies at high redshift in a color–color diagram. One color is obtained using two filters straddling the Lyman break. A second color is obtained from filters measuring the nonionizing continuum of the object. Thus, a star-forming galaxy at high redshift will be red in the blue color and blue in the red color. A non star-forming galaxy will also display a Lyman break but its nonionizing continuum will be red and the object could be confused with galaxies at lower redshift. For this reason, the Lyman-break technique is most powerful in identifying star-forming galaxies without large, pre-existing, older stellar populations. For these objects it can achieve success rates around 70 per cent or even higher.

An extensive review of the technique and of the properties of the galaxies thus selected is provided by Giavalisco [97]. Here, will we focus on aspects of its application to very high redshift galaxies.

The Lyman-break technique has been applied very succesfully both on the ground [259] and with the Hubble Space Telescope starting with the Hubble Deep Field [153, 295]. Originally the technique was seen as an efficient way of selecting candidates for spectroscopic followup. However, the early efforts verified that Lyman-break galaxies samples at redshifts of 3 and 4 were very robust and criteria could be found guaranteeing 70–80% of successful spectroscopic formation. On the basis of this success and partly because imaging, especially with the Hubble Space Telescope, was identifying samples too faint for extensive spectroscopic confirmation, Lyman-break galaxies samples are now often analyzed with the expectation that redshifts will not be measured for most of the objects. Strictly speaking, using Lyman-break galaxy samples 'blindly', without any spectroscopic confirmation, requires a leap of faith, as former success at redshift $z \lesssim 5$ is not a guarantee of performance at $z \gtrsim 6$. However, indications so far are that some selection criteria appear to be highly successful also at $z \sim 6$ [161].

6.2.1
The Lyman Break as a Function of Redshift

At low redshift there are only a small number of absorbers along the line of sight and a continuum break is found at a wavelength of 912 Å or lower when it is produced by absorbers at a significant lower than that of the galaxy under consideration. As the redshift increases, each absorber contributes its Lyman α absorption line, the Lyman α forest becomes denser and denser and the break moves toward 1216 Å. This is illustrated in Figure 6.1 where we show two synthetic rest-frame spectra of galaxies at $z = 3$ and $z = 5$, respectively. The spectra have been obtained using the synthetic stellar population models by Bruzual and Charlot [43] for a 100-Myr galaxy with solar metallicity undergoing star formation at a constant rate. The intergalactic absorption has been modeled following Madau [152].

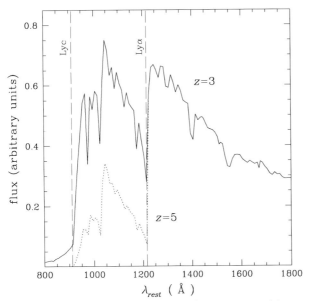

Fig. 6.1 Rest-frame spectra of two model galaxies at $z = 3$ (solid line) and at $z = 5$ (dotted line). It is clear that as the redshift increases the break effectively moves from the Lyman limit (vertical dashed line marked 'Lyc') to Lyman α (vertical dashed line marked 'Lyα').

A more thorough study would include also alternative descriptions of the IGM attenuation [164].

Clearly, especially at intermediate redshift, the effective wavelength of the break will vary from line of sight to line of sight. At high redshift these variations become less significant as the number of absorbers becomes large.

6.2.2
Synthetic Stellar Population Models

The availability of accurate synthetic models of stellar populations enables us to construct synthetic spectral libraries that can be used to fit the spectral-energy distribution (SED) and to derive the redshift of observed high-redshift galaxies. This is a cheaper alternative to spectroscopy, which is often inpractical or impossible for the faintest objects imaged today with, e.g., the Hubble Space Telescope.

Ideally, such a library of SEDs would allow for different star-formation histories, including simple stellar populations, multiple populations, constant star-formation rate systems, a range of metallicities and ages, dust content and a variety of extinction curves. Emission lines can be added self-consistently to each spectrum by computing their contribution from first principles or by interpolation. Knowledge of the luminosity function of the major galaxy types and its evolution can also be incorporated in the models. This needs to be done carefully as it might introduce a circularity into the argument when trying to determine the luminosity function and its evolution. Each SED would be subject to the mean effect of intergalactic extinction at the redshift being considered and then convolved with filter passbands to simulate an observation. The ultimate aim would be to create a tool that enables one to test possible selection criteria for high-redshift galaxies and estimate their effectiveness and the number of low-redshift interlopers to which they would be subject.

In the following we will use one such tool [185] to study a variety of selection criteria.

6.2.3
Redshift 6 Dropout Galaxies

Imaging in the visible has traditionally been superior, both in terms of depth and in terms of field of view, to imaging in the near-infrared. Thus, when the Lyman-break technique is applied to redshift 6 one is generally considering only one red filter for detection and rejection of the interlopers is done by excluding objects detected in the visible and blue bands. The Advanced Camera for Surveys (ACS) Wide-Field Camera (WFC) has been the premier imaging camera on the Hubble

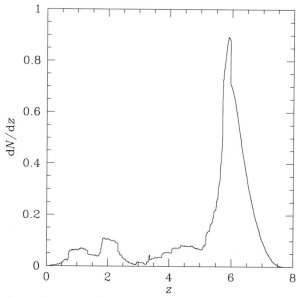

Fig. 6.2 Histogram of the expected redshift distribution of objects selected by the i-dropout criterion presented in the text. The histogram has been computed on the basis of a large number of synthetic spectral energy distributions [185]. According to this distribution, 24% of the objects selected by this criterion are at $z < 5$ and are therefore interlopers.

Space Telescope for several years and has been used to carry out both the Great Observatories Origins Deep Survey (GOODS [98]) and the Hubble Ultra Deep Field (HUDF [18]). The selection of redshift 6 galaxies in this case is carried out following the GOODS/HUDF criteria, namely:

$$i_{775} - z_{850} \geq 1.3 \tag{6.1}$$

$$S/N(z_{850})6 > 5 \tag{6.2}$$

$$S/N(V_{606}) < 2 \quad \text{or} \quad V_{606} - z_{850} > 2.8 \tag{6.3}$$

$$S/N(B_{435}) < 2 \tag{6.4}$$

where for B_{435}, V_{606}, i_{775}, and z_{850} we denote the magnitudes in the AB system [184] in the F435W, F606W, F775W, and F850LP ACS/WFC filters, respectively.

Deep imaging from the ground on 8 m class telescopes can reach depths comparable to that of GOODS [307] (without a comparable morphological information) but, for studying objects at the HUDF level, HST remains unrivaled.

In Figure 6.2 we show the redshift distribution obtained at the depth of the HUDF for i-dropout galaxies selected according to the criterion given by (6.1)–(6.4). The low fraction of low-redshift interlopers predicted here (~ 24%) is in agreement with the results of deep spectroscopic observations [161].

6.2.4
Lyman-Break Galaxies at Redshift Greater than 6

For redshift greater than 6, the Lyman break moves to the infrared and deep infrared data are required in combination of deep visible data. Up until Hubble Servicing Mission 4, installing the Wide-Field Planetary Camera 3 on HST (see Chapter 9), the premier near-infrared imager was the NICMOS instrument. Imaging data with NICMOS are best obtained with the F110W (J_{110}) and the F160W (H_{160}) filters.

Objects at redshift 7 or higher could be found using the criteria [33]

$$z_{850} - J_{110} > 1.3 \tag{6.5}$$

$$z_{850} - J_{110} > 1.3 + 0.4(J_{110} - H_{160}) \tag{6.6}$$

$$J_{110} - H_{160} < 1.2 \tag{6.7}$$

$$S/N(H_{160}) > 4.5 \quad \text{and} \quad S/N(J_{110}) > 2 \tag{6.8}$$

$$S/N(V_{606}) < 2 \quad \text{and} \quad S/N(i_{775}) < 2 \tag{6.9}$$

where J_{110} and H_{160} indicate AB magnitudes in the NICMOS filters F110W and F160W, respectively. In Figure 6.3 we show the histogram of the redshift distribution of Lyman-break galaxies selected according to the criteria given by (6.5)–(6.9).

The installation of the Wide-Field Camera 3 on HST will enable improved searches thanks to its higher sensitivity, larger field of view and improved filter set. A pos-

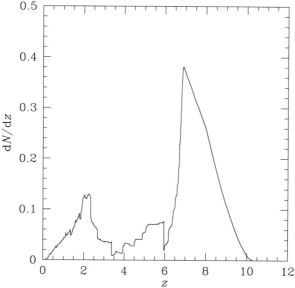

Fig. 6.3 Histogram of the redshift distribution of objects selected by the F850LP-dropout criterion presented in the text. The histogram has been computed on the basis of a large number of synthetic spectral energy distributions. The prediction is that 29% of objects selected by this criterion are interlopers at $z < 6$.

sible selection criterion for galaxies at $z > 6$ is the following:

$$z_{098} - J_{125} > 1.8 \tag{6.10}$$

$$J_{125} - H_{160} < 0.02 + 0.15(z_{098} - J_{125} - 1.8) \tag{6.11}$$

$$V_{606} - J_{125} > 2.4 \tag{6.12}$$

$$S/N(J_{125}) > 10 \tag{6.13}$$

$$S/N(V_{606}) < 2 \tag{6.14}$$

where V_{606}, z_{098}, J_{110}, and H_{160} represent here AB magnitudes in the WFC3 filters F606W, F098M, F125W, and F160W, respectively. In Figure 6.4 we show the histogram of the redshift distribution of Lyman-break galaxies selected according to the criteria given by (6.10)–(6.14).

Regardless of the specific criterion adopted, the important lesson here is that detailed modeling is needed to optimize a search technique. An important component of this modeling is the impact of photometric errors that could scatter low-redshift galaxies into the selection window for high-z galaxies. The opposite effect of scattering high-z galaxies out of the selection box is also possible. In Figure 6.5 we show the results of Monte Carlo simulations run on the HUDF catalog. The figure is based on the i-dropout selection given in (6.1)–(6.4). Noise is added to the data in order to simulate shallower observations with different exposure time balance as in the real HUDF. This simulation shows that for a hy-

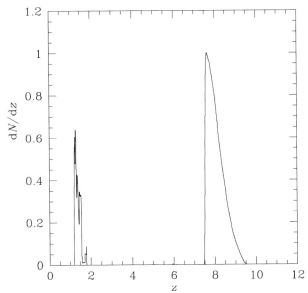

Fig. 6.4 Histogram of the redshift distribution of objects selected by the F098M-dropout criterion presented in the text. The histogram has been computed on the basis of a large number of synthetic spectral energy distributions. The prediction is that 15% of the objects selected by this criterion are interlopers at $z < 6$.

pothetical survey with half the integration time of the HUDF, the HUDF balance of equal exposure times in F775W and F850LP would be a good compromise between interloper contamination and completeness. Simulations such as these can be used to optimize the exposure times when planning a search for high-z galaxies.

6.3
The Lyman α Excess Technique

The Lyman α line can be very strong, especially in galaxies at low metallicity. Studies of Lyman-break galaxies indicate that about 30 per cent of LBG have Lyman α in emission, while the rest have Lyman α in absorption or no significant Lyman α [257]. However, some galaxies can be identified in a Lyman α search without being identified with a Lyman-break criterion either because their continuum is simply too faint or because it is intrinsically red.

Objects with strong Lyman α emission can be identified with the 'narrow-band excess' method, using either a set of narrow-band filters or a narrow-band filter and a broadband filter for continuum subtraction. The first technique is more demanding in terms of telescope time but it is superior since it enables searches at multiple wavelengths and it also allows, at least in principle, for a better

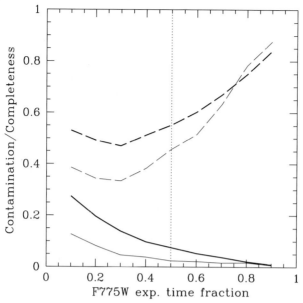

Fig. 6.5 Results of Monte Carlo simulations based on the HUDF and simulating observations of i_{775}-dropouts with about half (thick lines) or a quarter (thin lines) the total exposure time of the UDF. The x-axis gives the fraction of the integration time in the longest filters that is allocated to F775W. The y-axis shows the interloper fraction (solid line) or the incompleteness (dashed lines). As the detection band becomes deeper the interloper contamination is reduced but the incompleteness increases because the dropout band becomes too shallow. The vertical dotted line represents the HUDF solution of spending equal amounts of time in the dropout and in the selection bands.

continuum subtraction, minimizing the number of spurious sources [224]. An alternative method to search for emission lines objects is to do a slitless grism survey [161]. A variant of this method is the 'picket fence' method where one uses a number of very broad slits, giving up area in order to reduce the background level. The use of a narrow-band filter as a band-limiting filter reduces the background even further, in exchange for a more limited redshift-selection window [276].

6.4
The Balmer-Jump Technique

The Balmer-jump technique is similar to the Lyman-break technique but it relies on the 4000 Å break to identify galaxies at high redshift. The Balmer break arises only in stellar populations older than about 1 Gyr and therefore it cannot be directly used to identify galaxies at redshift greater than 6. However, the identification of a strong Balmer jump in a galaxy at lower redshift may be indirect evidence for very active star formation early on. More importantly, high-redshift z-dropout or

J-dropout galaxies samples obtained by means of the Lyman-break technique may well be contaminated by low-redshift galaxies with older stellar populations. The fact that galaxies with older populations are less likely to have strong emission lines will make it hard to distinguish spectroscopically between a high-redshift Lyman break – where Lyman α can be suppressed – and a low-redshift Balmer break – where line emission can be weaker or absent.

6.4.1
An Old Galaxy at Low or High Redshift?

An example of an object for which discrimination between a low-redshift Balmer break or a high-redshift Lyman break is complex is given in Figure 6.6, where we show images in various bands of an object identified in the Hubble Ultra Deep Field [172]. Spectroscopy of this object with the 10-m Keck telescope and the 8-m ESO/VLT has been inconclusive. Extensive modeling of the spectral energy distribution of this object indicates that the high-redshift solution ($z > 6.5$) is the most likely but a low-redshift solution, corresponding to an old, slightly reddened population, is also possible. Clearly, such a conclusion is to some extent dependent on the synthetic stellar population models used and, to some extent, on the priors adopted on the probabilities of the two solutions.

The published analysis gives a much higher probability for the high-redshift solution. From the published photometry of this object we can determine the likely

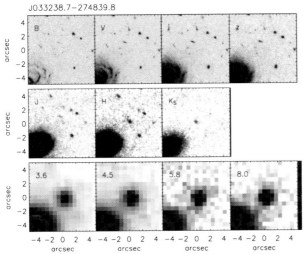

Fig. 6.6 Images of the HUDF galaxy J033238.7-274839.8 with the ACS/WFC HUDF (top row), NICMOS and ESO/VLT ISAAC (intermediate row), and Spitzer IRAC (bottom row). For the HST and ESO/VLT images the filter is indicated in the image, while for Spitzer IRAC the wavelength in micrometers is indicated. The galaxy is detected only in the H band and at longer wavelength. It is also detected by MIPS at 24 μm, is marginally detected in the J band and is not detected by ACS/WFC. If this object is a Lyman-break galaxy it is at a redshift greater than 6.5 [172] (Reproduced by permission of the AAS).

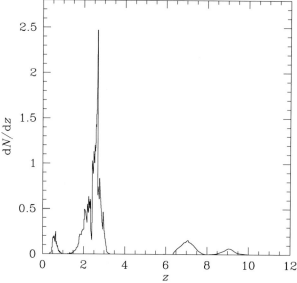

Fig. 6.7 Expected redshift distribution for the HUDF galaxy J033238.7-274839.8 using a luminosity function weighting that makes a low-redshift solution much more likely than a high-redshift one. This analysis would suggest that the object is likely to be at low redshift. The histogram is produced by considering more than 3×10^6 SEDs with a redshift spacing of $\delta z = 0.01$.

redshift distribution using the same SED library used for the Lyman-break criteria in this chapter. This is illustrated in Figure 6.7. In contrast to the published analysis, here the low-redshift solution appears more likely. These two diametrically opposite conclusions are derived from the same data and a similar analysis. The main difference in the second analysis is that it includes a luminosity function weighting implying that it is a priori much more likely to find a faint galaxy at low redshift than a bright galaxy at high redshift. It is this factor that drives most of the difference (see also Figure 6.9).

The same type of degeneracy is present for all Lyman-break selection at high redshift ($z > 6$) and suggests that examples of extreme galaxies at very high redshift should be taken with caution unless a spectroscopic confirmation is possible. The histogram of Figure 6.7 still leaves a probability in excess of 15% that J033238.7-274839.8 is at a redshift greater than 6.5. Thus, it is clear that on the basis of a spectral redshift distribution alone, it is very difficult to either confirm or reject the high-redshift solution for a galaxy of this nature. In the specific case a better argument in favor of the low-redshift solution for this object can be based on the non-observation of lower-mass galaxies expected on the basis of clustering arguments [180].

6.5
Photometric Redshifts

Photometric redshifts have matured in the last ten years and have become a powerful tool for estimating the redshift of galaxies. The technique is equivalent to using a very low resolution spectrum and it is based on the comparison of a set of colors with the colors of suitable templates at various redshifts. Generally, the larger the number of colors available, the more reliable the photometric redshift that is derived. Various groups have further improved the concept by adding priors based on the galaxy luminosity so as to exclude spurious solutions corresponding to galaxies at unrealistically high or low luminosity. Generally, the photometric redshift is most reliable when major spectral features like the Lyman break or the Balmer jump are straddled by the available colors. However, when enough colors are available, the technique becomes generally accurate up to an accuracy of 3 per cent in redshift [173].

There are two main classes of photometric redshifts, those based on observed templates and those based on synthetic models. Two other important factors are the number of templates and the option of including some form of luminosity weighting. The number of templates plays an important role but can also be misleading. Using few templates will generally mean that only one of the templates

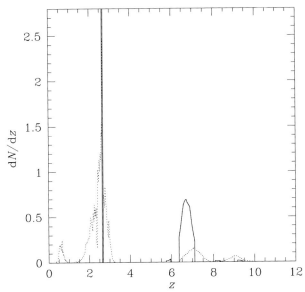

Fig. 6.8 Effect of the number of templates on the expected redshift distribution for the HUDF galaxy J033238.7-274839.8. The thin dotted line is the same as in Figure 6.7, while the solid line is for a subset of only 2.4×10^5 SEDs, which excludes models with dust and two component models. The high-redshift solution has now roughly the same likelihood as the low-redshift solution and both peaks are much narrower. The sharp low-redshift peak is offscale and represents galaxies with strong emission lines contributing to the broadband magnitudes in a very specific redshift interval.

will provide an acceptable fit to an observed SED and it will do so in a restricted redshift interval. The photometric redshift derived with this approach will formally have a small error bar. Adding extra templates with similar properties will provide an alternative fitting solution and will generally broaden the best redshift peak leading formally to a photometric redshift that is less good. This is illustrated for a specific galaxy in Figure 6.8. On the surface it would appear that this was a situation were extra effort leads to worse results but this is simply a consequence of the difficulty of estimating redshifts from broadband photometry. Clearly, which method is the best depends on the situation. For instance, if one is studying a range of redshift for which many spectroscopic redshifts are known, it will be possible to produce a template by stacking together observed spectra. This template will be very accurate and the high formal accuracy of the method will be also its real intrinsic accuracy for most objects. However, the same method applied, e.g., to a less well known redshift regime where the SED of galaxies are less well known will only give the illusion of high accuracy and might even introduce biases. In contrast, a photometric redshift method based on a rich library of observed or, more likely, synthetic spectral templates will tend to provide worse error bars. The derived redshifts and related error bars are the most conservative to use for a redshift interval with few known spectra but – by including SEDs that are rare or not observed – it will be pessimistic and suboptimal in a well-studied

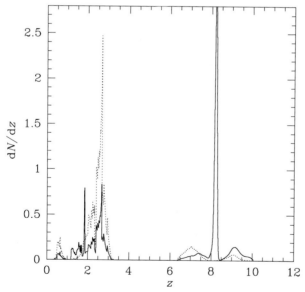

Fig. 6.9 Effect of luminosity function weighting on the expected redshift distribution for the HUDF galaxy J033238.7-274839.8. The thin dotted line is the same as in Figure 6.7 and includes luminosity function weighting while the solid line does not. The same SED library was used for both. The high-redshift solution is now most likely and the peak is much narrower. Note that the most likely redshift corresponds to zero probability in the luminosity function weighted case because it corresponds to an extremely bright and reddened object that is ruled out when a luminosity prior is used.

redshift interval where a targeted template set will provide much more accurate results.

Very often, photometric redshift distributions will have more than one peak, corresponding to radically different SEDs and redshift solutions. As we have discussed in Section 6.4.1 introducing a luminosity weighing prior can drastically change which redshift peak is selected, leading to two very different solutions. To illustrate this effect, we derive for our test galaxy the redshift distribution after suppressing luminosity function weighting and we show the result in Figure 6.9. The resulting redshift distribution is radically different from that obtained for the same SED library using luminosity function weighting and it is now compatible with the published value [172]. Clearly the decision of whether or not to use luminosity function weighting needs to be taken carefully as it can completely drive the result.

6.6 Samples of High-Redshift Galaxies

This section is devoted to a review of a few sample of high-z objects and to their implications for reionization.

6.6.1 Lyman-Break Galaxies at $z = 6$

The combination of the GOODS survey [98] and the HUDF [18] has created a sample able to begin probing the galaxy population at the reionization epoch. Galaxies

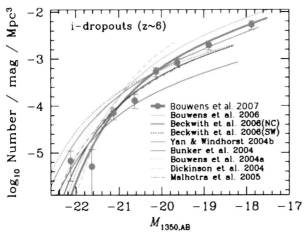

Fig. 6.10 The figure compares a number of different determinations of the luminosity functions $z = 6$. It is clear that once a few more discrepant determinations are excluded there is broad agreement on the resulting luminosity function (Reproduced by permission of the AAS).

at redshift 6 are selected in these surveys following criteria like those of (6.1)–(6.4). The catalogs derived by the various groups [32, 34, 44, 265, 306] are not in perfect agreement on an object-by-object basis but the overall picture that emerges is generally consistent (see Figure 6.10). Compared to the luminosity function of galaxies at lower redshift, there is some evidence for evolution. Different groups have argued for evolution in the value of M_* [34] or in normalization [18]. From a theoretical point of view one would probably expect both, as the number of halos at the mass scales probed by our the HUDF observations shows strong redshift evolution in numerical simulations. At the same time galaxies with the same halo mass a different redshift should be expected to have different star-formation histories.

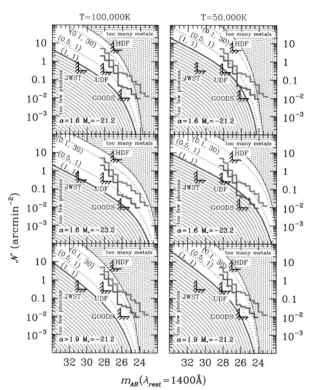

Fig. 6.11 Model predictions for Lyman-break galaxies developed in Chapter 4 compared to the data for the UDF and GOODS surveys. The left panels refer to very metal-poor, top-heavy stellar populations (e.g. Population III stars) while the right panels refer to top-heavy populations with metallicity typical of Population II stars, i.e. an effective temperature of 5×10^4 K. The lower and upper lines have been obtained for two different samples of redshift 6 candidate objects from the combined UDF and GOODS. The criterion corresponding to the lower curve being more robust but also more conservative. It is clear that if the galaxies we detect at $z = 6$ were at very low metallicity and thus at very high effective temperatures they would be able to reionize hydrogen (left panels). On the other hand, if they are enriched in metals they can only reionize hydrogen for the most optimistic, and least realistic, cases unless their luminosity function is very steep (see lower right panel) (Reproduced by permission of the AAS).

Figure 6.11 shows the expected cumulative surface density of reionization sources as a function of their apparent AB magnitude in the nonionizing UV continuum at a rest-frame wavelength of 1400 Å. Here, we have assumed that the comoving volume density of the sources is constant over the range of redshifts $5.8 \lesssim z \lesssim 6.7$ spanned by i-band dropouts in the UDF and GOODS [265]. The luminosity function of the sources is assumed to have the Schechter form, parameterized by its knee M_* and slope α. For reference, the Lyman-break galaxies at $z = 3$ have $M_{*,1400} = -21.2$ and $\alpha = 1.6$ [256, 305]. These are the parameters adopted for the top panels of Figure 6.11. The middle panels have a brighter knee ($M_{*,1400} = -23.2$), and the bottom panels have a steeper slope ($\alpha = 1.9$). The predictions in the left panels are for Population III stars; those in the right panels are for Population II stars. The models shown in Figure 6.11 are similar to those in Figures 4.6 and 4.7.

The requirement that the sources be able to ionize all the hydrogen in the IGM corresponds, for a given spectral shape, to a definite mean surface brightness at any chosen wavelength ($\lambda_{rest} = 1400$ Å in all cases considered here). This surface brightness depends on the fraction f_c of Lyman-continuum photons that escape from the sources and on the clumpiness C_ϱ of the IGM, which in turn determines the recombination rate. The surface brightness required for reionization fixes the normalization of the curves in Figure 6.11, which are labeled by the corresponding parameters (f_c, C_ϱ). These curves include both hydrogen and helium and the effect of recombinations (as described in Chapter 4).

Figure 6.11 shows that the detected i-dropout objects are not sufficient to reionize hydrogen unless the metallicity is very low ($< 10^{-3} Z_\odot$) or primordial or unless their luminosity function is very steep [306]. If the galaxies of redshift 6 are not sufficient to complete the reionization of hydrogen then it must be galaxies at higher redshift than do it. We will consider this possibility in Section 6.6.4.

6.6.2
Lyman-Break Galaxies at $z > 7$

Galaxies at redshift $z > 7$ can be identified on fields with deep near-infrared imaging. In practice, this is best done on fields imaged by HST NICMOS using selection criteria similar to those in (6.5)–(6.9). It is important to realize that the fraction of low-z contamination will depend critically on the depth of the sample. As an example, in Figure 6.12 we show the predicted contamination fraction when the selection given in (6.10) and (6.11) is applied down to $J_{125} = 26.5$ or only to $J_{125} = 24.5$. This change in the magnitude limit of the survey, changes the predicted contamination fraction from 15% to 97%. Using restrictive criteria able to minimize the number of contaminants is essential since the possibility of spectroscopically confirming any candidate is remote because of their faintness. Small samples of galaxies at this redshift have been identified [33, 186].

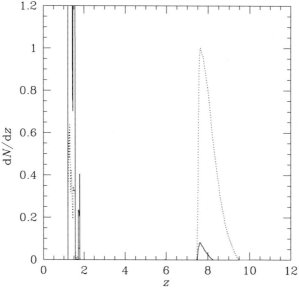

Fig. 6.12 Predicted redshift histogram for F098W-dropout objects selected as described in the text in a survey with magnitude limit $J_{125} = 26.5$ (dotted line) or $J_{125} = 24.5$ (solid line). The corresponding fraction of low-redshift interlopers changes from 15% to 97%. Note that the low-redshift peak in the $J_{125} = 24.5$ case is offscale.

6.6.3
Lyman α Emitters

As we have seen in Chapter 5, the evolution of the luminosity function of Lyman α emitters is often used to constrain the epoch of reionization. However, these galaxies are interesting sources of ionizing radiation. The ionizing photon output of these galaxies is proportional to the ionizing photons escape fraction f_c, while the Lyman α luminosity is proportional to $(1 - f_c)$. A very luminous Lyman α galaxy could be characterized by very low f_c and this would make it very ineffective for reionizing hydrogen. However, detecting their Lyman α emission allows us to place a lower limit to their total ionizing photon output and this is a more direct estimate than assuming a spectral energy distribution and extrapolating from the nonionizing continuum.

For an assumed value of the escape fraction f_c, the ionizing photon output of these galaxies can be derived directly and thus estimating their contribution to cosmic reionization is also more direct than for Lyman-break galaxies where in addition to f_c one also needs to assume an SED. In Figure 6.14 we compare the densities of Lyman α sources found in surveys searching for such galaxies at $z = 5.7$ and $z = 6.5$ to the model predictions of Chapter 4. The observations tend to be low compared to the models. The models were constructed for the primordial metallicity case and they could fit the data for higher values of f_c (thus making the predicted

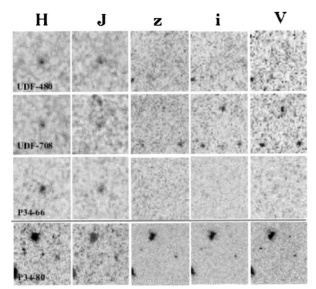

Fig. 6.13 Images in the V, i, z, J, and H bands (from right to left) for four candidates at $z \geq 7$ discovered in the UDF [186]. The last object (first from the bottom) has a S/N of only 5 and simulations suggest that there is a good probability of identifying similar spurious candidates where no object exists. (Reproduced by permission of the AAS)

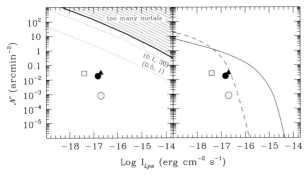

Fig. 6.14 Model predictions for Lyman α emitters developed in Chapter 4 compared to the data from a number of ground-based surveys. The models in the figure are identical to those in Figure 4.9. The left panel shows the surface number density for identical sources as a function of observed Lyman α flux. The right panel shows two luminosity functions. The solid line corresponds to $\alpha = 1.6$ and $M_* = -21.3$, while the dashed line corresponds to $\alpha = 1.1$ and $M_* = -17.5$. The filled symbols correspond to surveys at $z = 5.7$, while the open symbols correspond to $z = 6.5$. The circles are from the LALA survey [224, 225], the filled triangle is from Hu et al. [116] and the open square is from Taniguchi et al. [270]. The galaxies identified in these surveys are barely producing enough ionizing continuum to reionize hydrogen for $f = 0.5$ and $C_\varrho = 1$ if their luminosity function is characterized by a small value of M_*.

lines weaker) or for a mass function with a low value of $M_\star$. Adopting a nonprimordial metallicity would not drastically change this prediction because the increase in intrinsic strength of the Lyman α line for a primordial metallicity object is tied to a similar increase in ionizing continuum. Thus, while the stellar mass of the individual sources needs to increase if the metallicity is not primordial, the placement of the data points in Figure 6.14 does not change and the position of the various lines, which depends on the assumed ionizing photon output, changes by less than a factor 2, mostly due to the collisional contribution to Lyman α emission expected to be larger at lower metallicities.

6.6.4
High-Redshift QSOs

We have seen in Section 4.2.4 that once secondary ionizations are taken into account, QSOs have a minimum surface brightness between ~0.2 magnitudes brighter and 0.6 magnitude fainter than that of Population III stars. From Table 4.1 we find that for Population III stars the minimum surface brightness for reionization is ~29 mag arcmin^{-2}. Thus, QSO would need to have a surface brightness of at least ~28.8 mag arcmin^{-2} or fainter in their rest-frame 1400 Å continuum

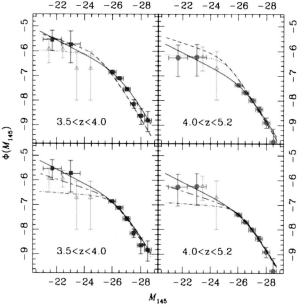

Fig. 6.15 Luminosity function of QSOs at $z = 3.5$–4.0 (left panels) or $z = 4$–5.2 (right panels). The squares are the combined LF data from the combination of SDSS at the bright end and GOODS at the faint end [87]. The triangles are from COMBO17 [299]. The upper and lower panels different in the type of models [64, 226] plotted (with permission from AA Copyright (2007) ESO).

in order to be able to reionize all neutral hydrogen in the Universe. The surface brightness of all 19 SDSS QSOs at $z \geq 5.7$ [80] is ~35 mag arcmin^{-2}, given that they are distributed over 6600 square degrees of area. We show in Figure 6.15 the luminosity function for QSOs at $z = 3.5$–5.2, which shows a knee at an absolute magnitude fainter than $M \simeq -26$. The QSO LF at $z \approx 6$ is a power law with a slope of -3.2 [79]. In order to reionize the Universe, this slope would need to be maintained from $M \simeq -26.5$ down to $M \simeq -24$. Thus, QSOs could in principle reionize hydrogen if the luminosity function measured for bright QSOs at $z \approx 6$ could be extrapolated all the way to faint magnitudes. This is possible but unlikely given that at lower redshift QSOs have a knee in their luminosity function at $M \simeq -26$ and a faint end slope flatter than the bright end slope. Despite this, it would seem that measuring the faint end LF of $z \approx 6$ QSOs should be a high priority.

If QSOs reionize hydrogen they would most likely also reionize helium at the same redshift and this – as we will discuss in Chapter 8 – might be problematic.

6.7
Fluctuations

If the first stars and the first galaxies are clustered, they will leave not only a signature in the near-infrared backgrounds (cf. Section 5.6) but will also contribute a fluctuation signal that might be easier to detect than a feature in the diffuse background spectrum. Fluctuations were detected in an analysis of the Spitzer data [126]. ACS GOODS data were used to eliminate the foreground galaxies. The cleaned Spitzer images reveal a correlation very different from that of the removed ACS galaxies, suggesting that the signal is not due to extended wings of the ACS galaxies (see Figure 6.16) and that the measured correlation might originate at

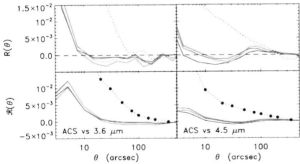

Fig. 6.16 The top panels of the figure show the measured dimensionless correlation $R(\theta)$ between the ACS B, V, i, and z bands (black, blue, green, red, respectively) and the Spitzer 3.6 μm (left) and 4.5 μm (right). The dotted line shows the detected correlation in the Spitzer bands. The lower panels show the integral of the fluctuation $(2/\theta)^2 \int_0^\theta R(\theta')\theta'\,d\theta'$ (Reproduced by permission of the AAS).

$z \gtrsim 6.5$. It is very plausible – although not yet universally accepted [273] – that these fluctuations are a real detection, however, their interpretation is still under discussion. An alternative explanation to the high-z interpretation is that the fluctuations are from dusty star-forming galaxies at lower redshift. Here, the fluctuation analysis would then display the same type of degeneracy as the one in Lyman-break selections and photometric redshifts discussed in Section 6.4.1.

6.8
Direct Detection of the First Stars

The direct detection of a Population III star is extremely challenging as they are individually very faint and also very rare. We can estimate their rarity with the same type of calculations that were used to obtain Figure 3.4 but obtaining instead a star-formation rate. The rate per comoving volume per unit redshift can be converted into a rate per unit area per unit redshift per year in the observer frame. This is shown in Figure 6.17. Within our model the maximum rate at $z \simeq 10$ remains as low as one star per year per 400 square degrees. This rate of formation would also roughly equal the rate of pair-instability supernovae resulting from these stars. The equivalence is only approximate because not all Population III stars end their

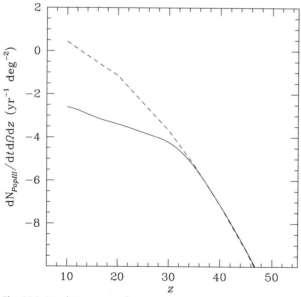

Fig. 6.17 Population III star-formation rate per square degree per year over a $\Delta z = 1$ for a model with Lyman–Werner suppression of Population III formation (solid line) or ignoring Lyman–Werner feedback (dashed line). The resulting rate in the case of feedback is very low.

life as pair-instability supernovae. The rate obtained ignoring LW-feedback is much higher [290].

Let us estimate the apparent magnitude of a Population III star. For a blackbody at 10^5 K the ratio between the bolometric luminosity, which is given approximately by the Eddington luminosity, and the value of νF_ν at 1400 Å is 0.096. For a 300$M_\odot$ star this gives us an absolute AB magnitude $M_{1400} \simeq -9$. This value may even be optimistic as, for the same mass, a detailed atmosphere model [39] yields $M \simeq -8$. The apparent magnitudes are $m_{1400} \simeq 38.5 - 40$ in the redshift range $z \simeq 10 - 25$. These are quite faint magnitudes. A Population III star would appear much brighter if we could observe it in the ionizing UV, but this is clearly not possible. Let us imagine that the Population III star is embedded in a dense neutral-hydrogen cloud. If we assume that two thirds of all ionizing photons produce a Lyman α photon the resulting Lyman α luminosity is 7.8×10^{39} erg s^{-1} or $(5.5 - 0.7) \times 10^{-21}$ erg s^{-1} cm^{-2} in the redshift range $z \simeq 10 - 25$. A pair instability supernova at $z \gtrsim 10$ would have a magnitude $AB \sim 25 - 27$ depending on the band and the redshift [290]. Finally, we have seen from (3.52) that a $10^8 M_\odot$ halo could form up to about 40 Population III stars. If all these stars formed almost simultaneously – which is unlikely at least according to the simulations – the luminosity of such a galaxy could be as high as $AB \sim 35$. As we saw in Chapter 5, the Lyman α cooling luminosity of such a halo is $\sim 3.3 \times 10^{-20}$ erg s^{-1} cm^{-2}. Such an object would have a continuum provided only by the two-photon processes and therefore enormous EW.

6.9
Hints for Further Study

- Assuming that the colors of Arp 220 are representative of a dusty starburst galaxy [246], determine how intrinsically fainter a galaxy with this spectral energy distribution would need to be at $z \simeq 2-2.5$ in order not to be detected by the ACS GOODS data or by the HUDF. Would these objects be detectable by IRAC or MIPS? Could these objects contaminate samples of massive evolved galaxies at high redshift [294]?
- Test how well we can predict completeness and contamination by deriving the i-dropout sample from the area of GOODS-CDFS covered by the HUDF. From the total catalog, one can generate a list of i-dropout candidates and random realizations obtained by changing flux according to the error bars to verify how many interlopers are on average added to the sample and how many real objects are lost. These results can be compared to the results obtained by deriving the i-dropout samples for the HUDF.
- Compare the effect of different models for intergalactic attenuation [152, 164] for a 100-Myr, Bruzual–Charlot model with solar metallicity in the $V_{606} - i_{775}$ versus $i_{775} - z_{850}$ color–color plot.

7
Deep Imaging and Spectroscopy Surveys

7.1
Overview

We have seen in the previous chapter that searches of high-redshift galaxies by imaging techniques play a major role in identifying the possible sources of reionization. To date, the deepest imaging surveys have been carried out with the Hubble Space Telescope. The first such public survey was the Hubble Deep Field (North, hereafter HDFN) [83, 295], that was followed by the Hubble Deep Field South [83, 296] and by the Hubble Ultra Deep Field [18] (hereafter HUDF, see Figure 7.1). The Great Observatories Origins Deep Survey (GOODS) [98] – while not strictly speaking a public survey – is also worth mentioning in this context because it reached a significant depth – comparable to that of the HDFs – over an area of about 320 arcmin^2, much larger than that of the deep and ultradeep surveys. All these surveys (including GOODS) released both their images and their catalogs to the public domain and triggered a number of follow-up observations.

In this chapter we will discuss the criteria for the choice of a particular field for a deep survey and then discuss the data-reduction steps required to analyze deep imaging data. Next, we will discuss how source catalogs are obtained. We will also consider potential surveys carried out with the help of a gravitational telescope and describe their limits and benefits. Finally, we will discuss options for deep spectroscopic surveys.

When an actual example needs to be presented, we will often refer to the Hubble Ultra Deep Field or more in general to HST instruments that enable the deepest imaging possible from the ultraviolet to the near-infrared (shortwards of 1.7 μm). The HST primer is a good source of high-level information on the Hubble Space Telescope and its instruments and it is generally reissued every year.

7.2
Field Choice for a Deep Imaging Survey

A number of considerations come into play when planning an ultradeep survey. Certainly a low galactic dust extinction is a strong constraint as well as the lack of

From First Light to Reionization. Massimo Stiavelli
Copyright © 2009 WILEY-VCH Verlag GmbH & Co. KGaA, Weinheim
ISBN: 978-3-527-40705-7

significant cyrrus if mid-IR observations are also planned. In addition to the specific requirements dictated by the combination of telescope and wavelength that one considers, there are often additional constraints due to other ancilliary observations that might be planned or might already by available. It is also worth observing that some of the criteria can have undesired consequence. For instance, the deficit of intermediate redshift elliptical galaxies in the HDFN is now tentatively attributed to the choice of avoiding radio sources in that field down to a faint limit.

As an example, for the selection of the Hubble Ultra Deep Field, a computer-based search was carried out to identify fields with the following characteristics:

- Low extinction and low cyrrus with the requirement of $E(B-V) < 0.02$.
- Low HI column density, with the requirement of $n_{HI} \leq 2 \times 10^{20}$.
- Absence of known bright sources: stars, galaxies, radio sources. For the latter objects the requirement was no radio source brighter than 10 mJy in the ACS field and no source brighter than 100 mJy within 10 arcmin.
- High ecliptic angle (> 30 degrees) to minimize zodiacal background.
- Sufficiently large area to maximize follow-up possibilities from the ground using large field of view instruments.

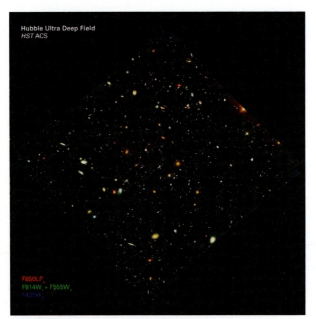

Fig. 7.1 False-color image of the Hubble Ultra Deep Field. The F435W filter was assigned to the blue channel, the F606W filter was assigned to the green channel, and the combination of the F775W and F850LP filter was assigned to the red channel. The image required 400 HST orbits, or about 1 million seconds on the sky and has an area of about 200 by 200 arcsec containing 10^4 galaxies (courtesy of S. Beckwith and HUDFteam).

7.2 Field Choice for a Deep Imaging Survey

The rejection of bright stars was based on the PPM catalog [228] containing 378910 stars. Galaxies were obtained from the RC3 catalog [68] including 23011 objects. QSOs were obtained from the Veron and Veron catalog [286] containing 30119 objects. Radio sources were extracted from the First catalog [292] containing 549707 sources. These numbers demonstrate the need for a computer-based search of suitable fields.

The fields obtained from the automated search were then screened for bright objects not included in the adopted catalog to obtain a final list of 13 possible fields. This was augmented by 12 already existing survey fields satisfying the $E(B-V)$ and n_{HI} conditions but not some of the other conditions. None of the fields from the automated search had extinction as low as the CDFS or the Lockman hole. Note that the CDFS itself was not selected by the automated search because of the presence of a nearby QSO.

Requiring accessibility to telescopes from both hemispheres eliminated several fields. The 5 surviving fields from the automated search were all located in the generic CDFS area (see Figure 7.2). The presence of a nearby QSO in CDFS did not seem to be an important disqualifying characteristic considering that it was the

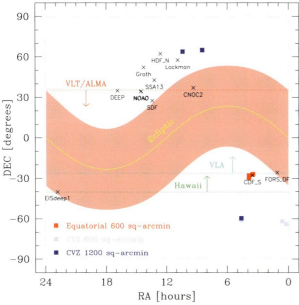

Fig. 7.2 Candidate fields for the Hubble Ultra Deep Field survey. Fields from the automated selection software are represented by filled squares. The red squares are fields with 600 sq. arcmin of viable follow-up field within ±30 degrees of the equator and thus accessible by observatories in both hemispheres. The light and dark blue are, respectively, fields in the continuous viewing zone of HST with 600 or 1200 sq. arcmin of viable follow-up field. None of them is in the equatorial band. The pink band represents the galactic zone of avoidance where crowding by stars and reddening become severe. The crosses represent existing ground-based surveys. Finally, the dotted lines represent the latitude limits of accessibility from Hawaii, from the VLA and from ALMA.

field with the lowest $E(B-V)$ of those visible from both hemispheres and had several pre-existing data sets already available including deep Chandra X-ray imaging. On this basis the CDFS area was selected for the HUDF.

The specific field within the CDFS was selected to avoid Chandra low exposure time areas, and to include a known $z = 5.8$ galaxy identified by GOODS [69]. As we will see, the fact that the HUDF is included in the GOODS-CDFS field is an added benefit because it allows a number of blind tests of our methods.

7.3
Observing Techniques for Deep Imaging Surveys

A deep survey tends to push the limit of what a particular telescope can do and therefore the observations must be carefully planned, executed, and analyzed. I will discuss below a few important steps.

7.3.1
General Considerations

HST is incapable of holding a given orientation for more than 50 days (at most). This implies that an observation like the HUDF lasting 400 HST orbits would be very challenging to do at one orientation. HST completes about 15 orbits per day but some of those are impacted by the South Atlantic Anomaly, an area of high radiation unusable for scientific observations. Completing the survey at one orientation would have required an average of 8 useful orbits per day, which is difficult to achieve and would have prevented any other HST science for a long period of time. Luckily, using just one orientation is not only difficult but also not desirable. Obtaining half the data at one orientation and the remaining half at an orientation differing by 90 degrees allows one to obtain two different versions of the same field with some of the detector-related artifacts being rotated and thus easier to remove. This is indeed the approach that was adopted to carry out the HUDF.

The images were obtained over a long time period, while still maintaining an angle between the telescope line of sight and the sun of at least 90 degrees in order to avoid the increased background that can be observed when this angle decreases below 90 degrees. This safe value was obtained by looking at a large number (~ 1000) of archival HST ACS/WFC images to determine the correlation between mean background and angle to the sun.

7.3.2
Dithering

Dithering consists in obtaining images of a field of interest changing the precise pointing by a few pixels between exposures. Dithering accomplishes two major goals: it cures the presence of bad pixels in a statistical sense by ensuring that an object will not fall on the same bad pixel on all images and it allows one to achieve higher photometric accuracy by smoothing over flat-fielding errors at the scale of the dithering steps. For undersampled cameras like the ACS/WFC onboard of HST, dithering can also be carried out with noninteger (subpixel) steps in order to improve the sampling of the point spread function (PSF) and obtain higher-quality images [128]. Precisely integer steps would in any case be difficult to obtain because of the pointing accuracy of HST and of the geometric distortion of the instruments.

For both CCD detectors and infrared detectors, having too many separate images tends to degrade the signal to noise as short exposures can turn a background-limited broadband image into a detector-limited image. Thus, for a given exposure time, the desire of having many dithering positions collides with the desire of maximizing the signal-to-noise ratio of the resulting combined image. For a very deep survey, the exposure time will generally be sufficient to allow good sampling at the subpixel level and a nicely spaced dithering pattern, while still obtaining sufficiently long background-limited images.

As an example, the dithering strategy for the HUDF was based on the following requirements:
- uniform sampling at the subpixel level
- coverage of the interchip gap of the ACS/WFC camera with exposure time equal to at least two thirds of the total exposure time
- avoiding placing a source on the same pixel multiple times

The dithering strategy that was finally adopted was to replicate with offsets a basic 4-point dither pattern sampling at the subpixel level. The offsets would be both along the interchip gap and across the interchip gap and no two exposures had the exact same pointing, in order to maximize the subpixel-level coverage and minimize the impact of bad pixels.

7.3.3
Super Bias and Super Dark

Biases and darks (or darks only for near-infrared data) need to be subtracted from each image before the images can be combined. However, doing so on an image-by-image basis would simply increase by a square root of two the detector-noise contribution. Thus, one generally combines multiple biases and darks. By combining hundreds of them, one can obtain the so-called super bias and super dark that will have particularly low noise. CCD detectors in space are vulnerable to radiation damage, especially in the form of generation of hot pixels, pixels with a particularly high dark current level. The CCDs of HST instruments are periodically warmed up to anneal some of the hot pixels (and by the way the temperatures are far too low to actually anneal the silicon crystal and it is unclear why the HST annealing process – which was discovered serendipitously – works, but it does). A super dark being averaged over a long period of time does not contain an accurate map of the hot pixels. For this reason, darks are obtained frequently and a daily or weekly dark can be obtained to mask out hot pixels from each individual image before the images are combined.

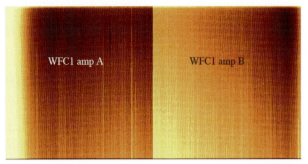

Fig. 7.3 ACS/WFC super bias for CCD WFC1 obtained by several individual bias frames [181]. The two quadrants have different mean bias levels. The vertical stripes are bad columns. The cross-hatching at a small angle from the horizontal direction is the herringbone artifact described in the text and is introduced by the lossless compression on the fly. Its regularity allows us to remove it by using a numerical filter (from Mutchler et al. [181], courtesy of STScI).

Biases and darks sometimes contain substructure or time-dependent effects such as bias jumps. These are features, generally in a band, on CCD images caused by sudden changes in the bias level. Since the bias represents the zero illumination level of the detector, these structures may need to be identified and removed.

More subtle effects sometimes need to be treated as well. For instance, the ACS/WFC instrument has the capacity of compressing data on the fly while the CCD is read out. The compression algorithm is based on a version of the Rice algorithm developed by Rick White and is lossless [261]. However, operating the instrument processor while the CCD is read out introduced a very low level electronic pattern in the data that appears as a herringbone effect (see Figure 7.3). Since biases and darks were compressed but the science images were not, the negative of this effect was imprinted on the final images by the bias and dark subtraction. This effect was discovered early [181] but not corrected because of its low level until Eddie Bergeron fully understood its nature and removed it through filtering [185].

7.3.4
Flat Fielding

When one attempts to observe objects that are significantly fainter than the sky background (zodiacal background for space observatories) one needs to achieve high levels of flat-fielding accuracy. The best way to do so is to rely on sky flats. For the typical high-latitude fields of interest for deep galaxy surveys, the fields are sparse enough that a flat field can be obtained from the sky images themselves. One danger with using sky flats is that if galaxies have faint extended and undetected halos producing a sky flat by combining many dithered sky images could lead to artifacts in the flat field. However, we find that, even at the depth of the HUDF, most of the sky is empty, so this is not a problem for imaging to the depth achievable today.

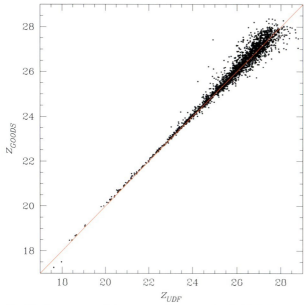

Fig. 7.4 Comparison of the magnitudes in z_{850} for the HUDF galaxies with the magnitudes of the same objects as measured in GOODS. Despite the ~ 2 magnitudes difference in depth between the two surveys galaxy magnitudes track very well each other down to the limit of GOODS. This suggests that galaxies do not generally have extended halos contributing significantly to their magnitude.

A test for this is to compare the magnitudes of the galaxies in the HUDF with the magnitudes for the same objects derived from the GOODS data, which are about 2 magnitudes less deep in the z_{850} band. This comparison is shown in Figure 7.4 and shows that magnitudes are very well correlated even for faint galaxies.

7.3.5
Image Combination

Drizzling is a powerful technique for combining dithered data originally developed for the Hubble Deep Field North by Fruchter and Hook [90]. Drizzling is capable of dealing with data with generic dither patterns and can be used to resample the point-spread function to improve measurements of objects shapes and photometry. The basic idea of drizzling is to define a grid of output pixels that can be of any size but, in the case of HST data, are often chosen to be smaller than the detector pixels. Each input pixel is *dropped* to the output pixel grid, contributing to each output pixel a fraction of the pixel flux proportional to the overlap in area between the 'drop' and each output pixel. This is illustrated in Figure 7.5. The algorithm is thus linear and flux conserving. Moreover, the output pixel grid can be corrected

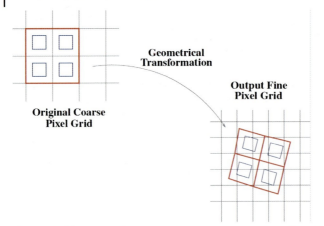

Fig. 7.5 An illustration of the drizzle algorithm [90]. An input area of 2 by 2 pixels (in red, left) is projected onto an output grid (right) with a smaller pixel size rotated with respect to the original pixel grid. Each input pixel is assigned a drop size smaller than its size (in blue). Note that in this example the central pixel in the output grid receives no flux information. Multiple dithered images will need to be drizzled in order to obtain an image without holes (with permission from PASP, ©(2002) by ASP).

for any geometric distortion present in the instrument so that the final image will be undistorted.

When dealing with data dithered at a subpixel level, drizzling is sometimes described as capable of *recovering the PSF quality* or even of *recovering the angular resolution*, while the former is broadly speaking correct, the latter is not.

Indeed, the PSF of every individual ACS/WFC image (for instance) can be thought of as the convolution of two terms:

$$PSF_{raw} = PSF_{optics} \star Pixel \tag{7.1}$$

where PSF_{optics} is the PSF that one would obtain for an infinitely small pixel size and *Pixel* is the top-hat function representing a pixel.

When drizzle is applied to this type of data, the resulting PSF is given by a further convolution with the Drop size kernel function, namely:

$$PSF_{raw} = (PSF_{optics} \star Pixel) \star Drop \tag{7.2}$$

Thus, strictly speaking, the combination by drizzle with any kernel other than the point kernel further degrades the image quality. In a situation where the detector pixel is very undersampled, as for instance in the case of the NICMOS instrument Camera 3 on HST, the optics PSF, PSF_{optics}, and possibly the dropsize, *Drop*, are negligible compared to the pixel size, *Pixel*, and the latter will be the limiting factor in angular resolution. Indeed, drizzling combination of well-dithered NICMOS camera 3 images to smaller pixel size yields final images that have a full width at half-maximum (FWHM) that is very close to the original pixel size, i.e. drizzle achieves the theoretical limit to the resolution of those images.

The only way to actually recover angular resolution would be to carry a deconvolution procedure [145]. However, deconvolutions are tricky [146, 147] and it is unclear that they would yield any substantial benefit when applied to HST data obtained after correction for spherical aberration.

Drizzling with generic parameters introduces a correlated noise component in the combined image. This is avoided when using drizzling with the point kernel but this is possible only when many tens of images are available, as it could otherwise lead to images with holes or very uneven exposure maps. The use of the point kernel is therefore restricted in practice to the deepest surveys. The amount of correlated noise introduced by drizzle can be characterized as a function of the drizzle parameters [48] as the relative reduction f_{rms} in noise for a single pixel compared to a large area and has been found to be:

$$f_{rms} = \begin{cases} \frac{s}{p}\left(1 - \frac{s}{3p}\right) & \text{if } s \leq p \\ 1 - \frac{p}{3s} & \text{if } p < s \end{cases} \tag{7.3}$$

where s is the scale of the output pixels compared to the input pixels and p is the drop size. When the point kernel is used $p = 0$ and (7.3) predicts no correlated noise. However, when analyzing in detail the noise properties of the HUDF we have discovered the presence of correlated noise even on images produced with the point kernel. Some of the extra noise was due to the herringbone artefact described in Section 7.3.3 and to an electronic ghost present in ACS/WFC and generating a negative image for each source in the image was also contributing. Even after these were eliminated, both the block average test and the autocorrelation test still revealed the presence of a correlated noise component. This additional component could be, in part, caused by a contribution from faint undetected sources but it is likely that a major contributor is a consequence of our overall strategy. Errors in the bias, dark and flat field are pixel dependent but when dithered images are combined these errors are applied to an extended area so that, for instance, a pixel with a positive error in the flat field will translate into a small area of the size of the dithering pattern characterized by a positive contribution. A fractional error of ε in the flat field will translate into a per pixel error of εf_{sky} if f_{sky} is the flux per pixel from the sky, this needs to be compared to the mean isophotal flux per pixel of the HUDF galaxies that is shown in Figure 7.6 and clusters around $z_{AB} = 33.3$. The sky brightness in the HUDF in z_{850} is of the order of $z_{AB} = 29$ per pixel so that the sky is ~ 4 magnitudes brighter per pixel than the HUDF galaxies. A flat-field error of 1% would then contribute significantly to the flux per pixel of typical faint HUDF galaxies.

While understanding the origin of the various correlated noise contributions is important, measuring them is a necessity for deriving good photometry and reliable errors for the galaxies in the image. An approach that seems to provide good results is to obtain rms maps from the variance maps produced by the drizzle software and rescale them so as to increase the effective pixel noise to the level that is relevant for galaxies under study.

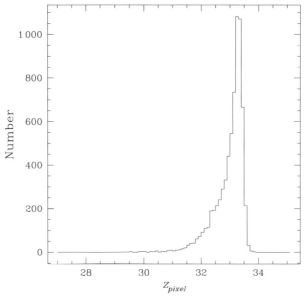

Fig. 7.6 Histogram of the magnitude per isophotal pixel in the HUDF galaxies. This is based on the flux within the isophotes used for galaxy detection. The mean magnitude per pixel for the full galaxy profile would be even lower.

7.4
Self-Calibration

We have so far focused on the standard approach in UV-optical-near-IR astronomy to remove instrumental signatures from imaging data. However, the alternative technique of self-calibration has been proposed [10, 86] and has been used, e.g., for Spitzer IRAC data and has been tested also on HST data. This technique is based on the idea of solving, iteratively but simultaneously, for the sky parameters and the instrument parameters. This is done by writing formal equations relating the observed signal to the instrument parameters and to the true sky image. Under the assumption that one is not too far from a solution, these equations can be linearized and, through a process that is essentially analogous to a sophisticated operator inversion, one can calibrate the instrument at the same time as the sky image is extracted. For complex instruments such as NICMOS on HST one can also add calibration images such as darks to the analysis and process them in the same fashion as the astronomical data. In principle, this method enables one to obtain not only a consistent calibration for the sky images but also accurate error estimates. In practice, this method is still not widely used. One of the difficulties being that it requires manipulation of large matrices, e.g., for NICMOS array with 256×256 pixels the full covariance matrix has 2.5×10^{11} elements. This matrix has a size dependent on the square of the total number of pixels, so for modern images with 4096×4096 pixels or larger it can reach sizes of $\sim 10^{14}$ elements.

Another difficulty is that the application of this technique to obtain final images with improved sampling from subsampled data has not been yet well documented in the literature.

7.5
Catalogs

Once the images have been combined, they can be used to derive source catalogs. Today, the most commonly used software for deriving catalogs is Sextractor [23] but the following considerations would be applicable to other software packages as well. It is important to realize that automated photometry of galaxies is intrinsically an ill-posed problem and is much more complex than stellar photometry. The basic difference lies in the fact that the shapes of galaxies cannot be assumed a priori and that their halos can be physical rather than being simply due to the point-spread function wings. Thus, deriving the total luminosity of a galaxy will generally entail an extrapolation under some set of assumptions. Such an extrapolation may well end up being different from band to band so that when measuring a color it will be important to ensure that the same isophote is used in all bands so as to compare equivalent physical portions of galaxies.

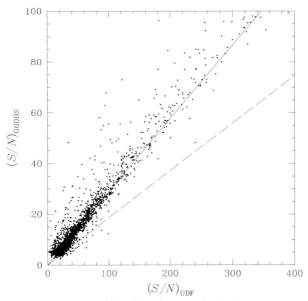

Fig. 7.7 Comparison of the achieved S/N in the F850LP image in GOODS and in the HUDF for the galaxies in common to both surveys. The dashed line represents the expect correlation based on the ratio of the exposure times. The solid line represents the expected correlation if the HUDF data are affected by an average contribution of correlated noise of 55%.

Estimating the photometric error and signal-to-noise ratio of a detection is also tricky as they will depend on the accuracy of the variance or rms maps and on the presence or absence of correlated noise. Unfortunately, for the deepest observations correlated noise cannot be entirely avoided as it can be imprinted on the data by the reference files even when adopting a point kernel as we have discussed in the prevous section. Therefore, one will need to rescale the variance or rms map by a factor dependent on the size of the objects of interest.

The amount of correlated noise can be estimated by block-averaging blank sky areas or from the power in the autocorrelation peak of cleaned images. Let us assume that we have an image and its related rms map, the next subsection will describe the major steps that are needed to carry out automated photometry on the image. In Figure 7.7 we show the detection S/N of galaxies as observed in the HUDF and in GOODS. The instrumental setup is the same and the two surveys differ only in their exposure times that for the F850LP filter shown in the figure are in a ratio of ~28.8. Unfortunately, as the figure shows, the gain in achieved S/N ratio is less than what could be inferred from the ratio of exposure times. The reason for this is that the HUDF when combined using the same method used for GOODS, reaches a systematic floor due to correlated noise effects. Some of these effects can be removed as discussed in Section 7.3.3, improving the S/N and approaching the expected values.

7.5.1
Layout of an Automated Photometry Algorithm

Finding objects in an image and measuring their photometry can be done in a number of logical steps. One good starting point could be to scan the images looking for the central parts of galaxies that are likely to have higher signal-to-noise ratio. One could in principle require to detect a single pixel with a S/N above some given threshold but this would make one vulnerable to detector defects that survived to the final image and would also be biased against more extended, lower surface brightness objects. A good compromise could then be to require a minimum number of pixels above some *noise* threshold providing together a minimum S/N for detection. This enables the detection of both extended objects requiring many pixels just above the minimum threshold in order to reach the detection S/N and compact sources reaching the detection S/N over just 1–2 pixels.

Once the galaxy cores have been identified, one can extend each of them by incorporating lower signal-to-noise areas below the original noise threshold but above some new *extension* threshold. Extending the area of the galaxy will provide for a better estimate of the galaxy flux.

Once areas associated to galaxies are identified, one will be able to obtain a sky map that can be used to estimate the sky underneath any individual galaxy. This could be done by interpolation or by appropriately fitting the sky image.

The galaxy detection and extension steps might have merged multiple galaxies into a single object. Depending on the specific extension step algorithms, it is also possible that galaxies with uneven luminosity distribution will be split into disjoint catalog objects. A necessary step is then to review the object list according to some preset criteria based on gradients, relative changes in luminosity, etc., in order to merge some objects and split some others.

A last step is then to carry out the photometry on the final list of galaxies. When more than one image is available, one could carry out the identification on one image (or a combination of the images) and carry out the photometry steps on each image so as to use the same definitions of objects in each band.

Summarizing the method presented here includes:
- Galaxy-core identification.
- Extension step to encompass whole galaxies.
- Sky-level determination.
- Merging and splitting of objects with substructure.
- Photometry.

This process is descriptive of, e.g., the steps of the Sextractor [23] automated photometry package. Of all the required steps, the merging and splitting of objects is the most ill-defined and it is possible that one could improve this step by considering images in different colors simultaneously, so as to also use color uniformity as an additional criterion for object merging. So far, no public software package makes use of this additional information.

An implicit assumption in the method described so far, is that the point-spread functions of the different color bands are comparable. If they are not, one common method is to convolve the narrowest images by a small kernel trying to degrade them to the same PSF width as the other images. When the PSFs are very different, this method is less desirable and in same cases it may be completely unacceptable. An alternative is to fit PSFs to the lower-resolution image using the positions of objects in the high-resolution image as a constraint. The TFIT package [133] developed by the GOODS team implements these concepts.

7.5.2
Sextractor Photometry Tips

Various software packages are available to carry out automated photometry but in recent years the software Sextractor [23] has become one of the most commonly used packages. Sextractor provides several different values for the magnitudes of an object, and while some choices are clearly inferior it is less clear what is the best solution. A solution that was found quite adequate for the HUDF program was to focus on the ISO and AUTO magnitudes and fluxes and select one band for detection, while carrying out photometry in the other bands in dual-image mode. This ensures that the area over which galaxy flux is measured in the various bands is the same, so that colors computed from these fluxes are meaningful. The best solution for colors is to compute them from the differences of ISO magnitudes, while a good estimate for the total galaxy magnitude is the AUTO magnitude. A good estimate for the detection signal-to-noise ratio of an object is the ratio of its ISO flux over its ISO flux error (see Figure 7.8). Note that in general the same quantity measured

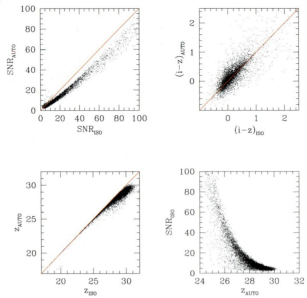

Fig. 7.8 Detailed comparison of ISO and AUTO quantities derived for the HUDF z_{850} image using Sextractor. The top left panel shows a comparison of the S/N ratio derived from the ISO and the AUTO fluxes, showing the trend for the AUTO S/N to be lower than the ISO one, which is the most relevant for detection. The bottom left panel shows a comparison between ISO and AUTO magnitudes. As expected, AUTO magnitudes are systematically brighter than ISO magnitudes. The bottom right panel illustrates a good way to characterize the depth of a survey, namely, the plot of the AUTO magnitudes versus the ISO S/N for the same object. This plot essentially gives the detection S/N for an object of a given total magnitude. Finally, the top right panel compares the $i_{775} - z_{850}$ AB colors for the HUDF galaxies. The plot shows a large scatter (although no obvious systematic effect) that is due to the fact that AUTO colors in two different bands refer to different spatial regions.

using the AUTO flux and its error is lower, reflecting the fact that in order to get a better estimate of the total magnitude, one has to also include lower flux pixels that decrease the overall S/N.

Clearly, given the residual degree of arbitrariness in these choices, it is essential that any specific method be validated using realistic Monte Carlo simulations so as to verify the photometry and its errors and identify possible biases that are generally known to be present.

7.5.3 Simulations

Given the complexity of the cataloging software and the ill-posed nature of the problem, simulations are of fundamental importance in order to estimate the noise level as a function of galaxy size and luminosity as well as to determine the presence of photometric biases and characterize them. These simulations are traditionally carried out through a Monte Carlo approach. Simulated galaxies of known magnitude are added to the images and recovered using the same Sextractor parameters as the real catalog. This enables one to determine the recovery probability as a function of the input magnitude and of the light profile shape and galaxy size. The dependence from the latter parameters introduces a model dependence in the simulations and one must be careful that the distribution of input parameters is compatible with the observed distribution. This is illustrated for the HUDF V-dropout galaxies [185] in Figure 7.9. The dependence of the unknown distribution of input parameters is the major disadvantage of simulations based on idealized galaxies. In addition to the recovery probability (or completeness) the simulations using idealized galaxies also allow one to derive photometric biases and true photometric errors since the true input magnitude is known for each object.

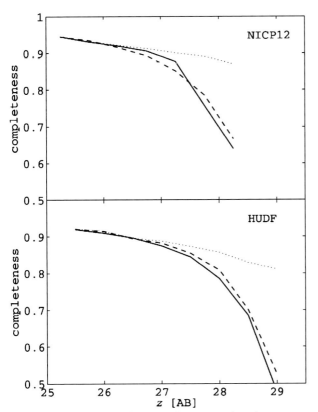

Fig. 7.9 Completeness for the HUDF (bottom panel) and one of the HUDF parallel fields (top panel). The solid lines refer to simulated galaxy simulations, while the dashed lines refer to simulations with replicated galaxies. The thin dotted line illustrates the effect of blending of sources. It is comforting that the two simulations are basically in agreement with one another (Reproduced by permission of the AAS)

An alternative approach is based on replicating and dimming real galaxies. This method bypasses the uncertainties related to the unknown distribution of galaxy properties but presents difficulties of another nature. If galaxies are dimmed only by a small amount before being added back to the image the sky surrounding and below each object will increase the noise level, altering the recovery probability as the dimmed object will be noisier than an object of the same magnitude in the original image. Dimming objects by a large factor overcomes this problem at the price of requiring a large extrapolation in properties of galaxies. Dimming of real galaxies by a large factor also implies only a small and easily correctible excess in noise for the simulated objects. Another disadvantage of using real galaxies is that their true magnitude is not really known so that computation of photometric biases is less well founded. An alternative to replicating individual galaxies is to add back to an image a complete rescaled rotated or shifted image. This has the advantage that local artifacts introduced by adding a small subimage for each galaxy are avoided but the disadvantage that the crowding properties of the image may be altered. In practical cases, as e.g., the HUDF, the latter concern was not a serious issue.

Given the pros and cons of simulated vs. dimmed objects the best approach may well be to carry out both and compare the results. Agreement between the two methods suggests that the recovery fractions derived are robust (see Figure 7.9).

7.6
Cosmic Variance

Galaxies are not uniformly distributed but tend to cluster in filaments or in clusters at the convergence of multiple filaments (see Figure 7.10). Studying a field through a realization of the high-redshift Universe may lead to large variations in the mean number of objects when one focuses on different fields (see Figure 7.10). These fluctuations are known as cosmic variance. Cosmic variance is particularly insidious for moderate sample sizes. Indeed, when only one or two galaxies per field are present, the Poisson fluctuations on this number will be dominant. In contrast, when one considers a sample with several objects per field, cosmic variance becomes very important as it causes statistical errors to behave differently from the square root of the number of candidates. Ignoring cosmic variance tends to overestimate the statistical significance of conclusions drawn from these samples.

The effect of cosmic variance can be studied either through tracing pencil beams across N-body simulations or analytically on the basis of the correlation function [248, 279].

7.7
The Gravitational Telescope

Gravitational lensing can amplify the luminosity of distant galaxies by very significant amounts. However, gravitational amplification conserves surface bright-

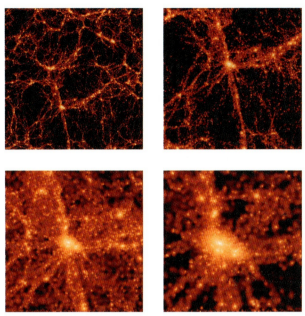

Fig. 7.10 Illustration of the formation of structure in a cosmological N-body simulation. Galaxies tend to reside in filaments and the distribution is far from uniform (courtesy of Joerg Colberg and the Virgo consortium [121]).

ness and an amplification A is achieved by increasing the object area by the same amount. Thus, the useful area in the source plane decreases by A. For galaxies with a luminosity function $\phi(L)$ the number of objects detected to a given flux at a given redshift, $n(L_{min}, z)$, is given by $(dV(z)/dz)\Phi(L_{min}, z)$, where $V(z)$ is the cosmic volume and $\Phi(L, z)$ is the cumulative distribution function that is given by:

$$\Phi(L, z) = \int_{L}^{\infty} \phi(L, z)\, dL \tag{7.4}$$

where $\phi(L, z)$ is the luminosity function. Thus, we can connect the number of objects lensed by A to the intrinsic number noticing that [37]:

$$n'(L_{min}, z) = A^{-1} \Phi(L_{min}/A, z) \frac{dV(z)}{dz} \simeq A^{\beta-1} n(z) \tag{7.5}$$

where β is the effective slope of the cumulative distribution function defined as:

$$\beta = -\frac{d\ln\Phi(L, z)}{d\ln L} \tag{7.6}$$

It is sometimes assumed that gravitation lensing is a powerful tool to study the faint end of the luminosity function of high-z galaxies. This is actually not the case. One reason is the limited effective area available at a given amplification. The resulting small-number statistics and the uncertainties in modeling the lens make the derivation of a robust luminosity function quite challenging. The other reason why studying the faint end of the luminosity function by lensing is problematic has to do with the mechanics of lensing and the shape of typical luminosity functions.

Let us consider for instance the luminosity function $\phi(L)$ of i-dropout galaxies [33] that has a faint end slope ≈ -1.7. The cumulative distribution function will have a faint-end effective slope $\beta \approx 0.7$ so that the number of lensed objects will be actually lower than the number of unlensed ones. More generally for any luminosity function $\phi(L)$ with a faint end slope less steep than -2, the decrease in effective area will not be compensated by the increase in depth [35]. In contrast, for luminosities above L_* the typical luminosity function of the Schechter type will be exponentially decreasing and therefore characterized by an effective slope β steeper than one. Thus, gravitational lensing is most effective in finding relatively bright objects in shallow searches over lensing clusters rather than in trying, e.g., to measure the faint end of a luminosity function by looking for very faint sources in deep searches [35].

Lensing may also enable followup study, for instance spectroscopic studies, of objects of luminosities too faint to be studied directly for the unlensed field samples. Apart from serendipitous findings, this can be very productive at faint mag-

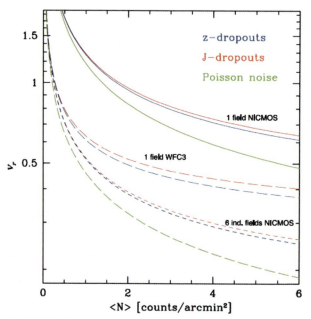

Fig. 7.11 The figure shows the total fractional error in the counts, including cosmic variance, as a function of the density of objects [279]. Each observing configuration consists of three lines, the top one is for J-dropouts, the intermediate one for z-dropouts and the lower one represents Poisson noise and is given for reference. The three sets of lines are for a single NICMOS Camera 3 field (solid lines), a single WFC3 field (long dashed lines) and 6 independent NICMOS Camera 3 fields (short dashed lines) (Reproduced by permission of the AAS).

nitudes where the number density of objects is high because even a decrease in the number of objects found still leads to an acceptable sample. For instance, if lens modeling of a given cluster shows a 0.5 square arcmin effective search area amplified by a factor 5 or more, the optimal depth to be targeted would be that enabling observations of objects with an intrinsic density of a few per square arcmin detectable at amplification 5 or more.

7.8
Deep Spectroscopy

HST is capable of reaching in imaging much fainter magnitudes than larger ground-based telescopes thanks to its low intrinsic background and to the sharper point-spread function (at least in the visible) resulting in even lower effective background (see discussion in Section 9.4.1). This background reduction is the key in broadband images that are background limited, i.e. are such that the main noise contribution in the image is the sky background B rather than, e.g., the detector noise D, or $B > D$.

In slit spectroscopy the signal for an astronomical object is dispersed into a number of bins proportional to the resolving power R and the background is similarly reduced to B/R, while the detector noise D remains the same. Thus, for sufficiently high resolving power R, the detector noise becomes dominant, $B/R < D$. This is why HST's performance in spectroscopy does not match its prowess in imaging. The degradation in performance of ground-based telescopes is less dramatic but it is still present. As a result, the faintest objects imaged by HST are beyond the present capabilities for a spectroscopic study.

7.8.1
Spectroscopic Analysis Techniques

While software like drizzle and Sextractor are very widely used in analyzing deep imaging data, no software for the reduction and analysis of deep multiobject spectroscopic data has reached a similar degree of prominence. There are probably several reasons for this. Spectroscopic data are more complex than imaging data and it is probably more difficult to develop software that is relatively independent of the specific instrument. Despite these difficulties, some of the steps are relatively standard and present challenges that would benefit from a standardized approach. As an example, wavelength calibration could be done with or without resampling the data. If done without resampling the data, a suitable description of the wavelength associated to each pixel will be needed and this might require a high-order polynomial or a more complex function in order to achieve the desired accuracy. On the other hand, if the data are resampled one would want to do so without adding unnecessary broadening to the spectra. Spectral extraction is another subtle problem as one would want to adopt an optimal extraction method. Unfortunately, for low S/N spectra of nonpoint-like sources like galaxies the profile shape is not known a priori and a profile obtained by collapsing in wavelength the spectrum may be affected by low S/N and possibly a wavelength dependence. It would appear that this type of problems are general enough and complex enough to warrant a systematic approach and it would be desirable to have access to easy-to-use software of general applicability for the extraction of multiobject spectra.

7.8.2
Slit Spectroscopy of Faint Targets

The difficulty in matching imaging depth with spectroscopy is illustrated in Figure 7.12 where we plot the magnitude versus S/N diagram for the galaxies in the HUDF and estimate the integration time on the FORS2 instrument on the ESO/VLT required to obtain a $S/N = 3$ in the continuum at 9000 Å. This S/N is the minimum for a detection of the continuum and enables detection of faint lines. About 90% of the HUDF galaxies are fainter than $z_{850} = 25.5$ and would require more than 100 h of integration and are therefore practically unfeasible. Unfortunately, the majority of the i-dropout galaxies falls in this category.

Improving on this mismatch between imaging and spectroscopy requires red optimized spectrographs, high multiplexing, and a field of view larger than a few

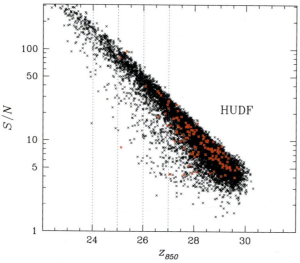

Fig. 7.12 Distribution of z_{850} magnitude versus S/N for the HUDF galaxies. The i-dropout galaxies are identified as red dots. The four vertical blue thin lines identified the magnitudes for which a $S/N = 3$ integration at 9000 Å is obtained with FORS2 on the ESO/VLT with integration times of 7 h, 46 h, 290 h or 1800 h, respectively from left to right. If we consider an integration time of 100 h as the maximum practical limit, we can see how objects fainter than $z_{850} \simeq 25.5$ are too faint for a spectroscopic study of their continuum or of their faint lines. Clearly, objects with strong emission lines are easier to study.

arcmin to be able to cover the whole HUDF in a single field. Ultimately, major improvements will be brought about by the next generation of telescopes.

7.8.3
Slitless Spectroscopy

In slitless spectroscopy a prism or a grism is inserted in the beam of an imager to obtain spectra for all objects in the field. Clearly, the sensitivity will be lower than in imaging or slit spectroscopy as the object signal is dispersed, while the background is fully present. Despite this limitation slitless spectroscopy is relatively powerful on HST as the background is intrinsically low [161].

7.9
Hints for Further Study

- Derive Sextractor catalogs for the z_{850} image of the HUDF using different extraction parameters, e.g.: *version 1* –
  - DETECT_MINAREA=16,
  - DETECT_THRESH=0.6,
  - ANALYSIS_THRESH=0.6,
  - THRESH_TYPE=RELATIVE,
  - DEBLEND_NTHRESH=32,
  - DEBLEND_MINCONT=0.03,
  - CLEAN=Y,
  - CLEAN_PARAM=1.0

  and *version 2* –
  - DETECT_MINAREA=10,
  - DETECT_THRESH=0.759,
  - ANALYSIS_THRESH=0.759,
  - THRESH_TYPE=RELATIVE,
  - DEBLEND_NTHRESH=8,
  - DEBLEND_MINCONT=0.03,
  - CLEAN=Y,
  - CLEAN_PARAM=0.5.

  How do these catalogs compare? Is there an obvious superior choice?
- Determine the correlated noise contribution by using block averages of empty sky areas on the HUDF on scales typical of faint galaxies ($z_{AB} = 29$). How large is their size and how large is the correlated noise contribution on those scales?
- Compare spectroscopic redshift for GOODS-CDFS [285] with photometric redshift from a public package [21, 36]. Investigate the nature of discrepancies.

8
The Reionization of Helium

8.1
Overview

The reionization of helium is an important phenomenon affecting our understanding of the evolution of the integalactic medium at high redshift. As we will see there is evidence for a late reionization of helium compared to hydrogen. This represents one of the strongest arguments to suggest that hydrogen was reionized by stars, as the harder spectrum of quasars would have reionized both helium and hydrogen at the same time. A similar argument could be made for reionization by Population III stars, unless one finds a way to recombine helium, at least partially, without recombining hydrogen. A late reionization of helium also produces a late reheating of the IGM. Thus, it has consequences on the IGM temperatures that we measure at low redshift so that the IGM temperature is an indirect probe of HeII reionization. In the following we will consider both direct and indirect indicators.

8.2
Gunn–Peterson Troughs in QSOs

Observation of HeII absorption is complicated by the fact that only ~15% of the lines of sight to $z \simeq 3$ are free of Lyman-limit systems so that there are less than ten known $z \simeq 3$ QSOs for which a helium Gunn–Peterson test can be attempted. Lyman-limit systems are QSO absorption systems with neutral-hydrogen column density of at least $n_{HI} \simeq 10^{17}$ cm^{-2}. They will produce a major step in absorption below 912 Å in their rest frame and even though the optical depth decreases with decreasing wavelength it still remains significant at 228 Å, which corresponds to the ionization energy of Helium II.

Another difficulty in undertaking this measurement is that the HeII Lyman α line is at 308 Å, which remains in the UV at $z \approx 3$ and it is thus accessible only from space. Indeed, progress in this field has been made mostly thanks for HST and FUSE [66,81,111,119,131,223,244,311]. These studies have shown an increase in the Gunn–Peterson optical depth of HeII as the redshift approaches $z \simeq 3.0$. In Figure 8.1 we show a spectrum of He 1700 + 6416 at $z = 2.72$ [81].

From First Light to Reionization. Massimo Stiavelli
Copyright © 2009 WILEY-VCH Verlag GmbH & Co. KGaA, Weinheim
ISBN: 978-3-527-40705-7

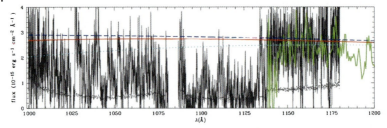

Fig. 8.1 Fuse (black solid line) and HST/STIS (green line) spectrum of HS 1700+6416 at $z = 2.72$ [81]. The solid red, dashed blue and dotted cyan lines are the extrapolated continuum using different extinction curves. The large gap at 1085 Å is in correspondence with a gap in the two FUSE detector arrays. The spectrum shows a substantial decrease, in correspondence of the Lyman α for helium III (with permission from AA, Copyright (2006) ESO).

In Figure 8.2 we show HeII Gunn–Peterson measurements from 6 QSOs supporting a rapid increase in optical depth at $z \gtrsim 2.8$. Similarly to our discussion in Chapter 5 for the Gunn–Peterson effect for hydrogen, even in the case of helium, He II in neutral Lyman α systems could mimic the presence of a diffuse medium [151]. Moreover, an additional difficulty in studying the HeII reionization is that at redshift $z > 3$ hydrogen Lyman α absorption at low redshift can mimic He II absorption at high redshift. As for the case of hydrogen the dumping wing predicted to characterize Gunn–Peterson troughs would make the interpretation simpler, but unfortunately it is generally not observed. SDSS J1711+6052 at $z \simeq 3.82$ is one of the notable exceptions (see Figure 8.3) although the precise interpretation of the observations is still unclear [312]. Despite these problems there is now growing consensus that HeII reionization was completed at $z \approx 3$–3.4.

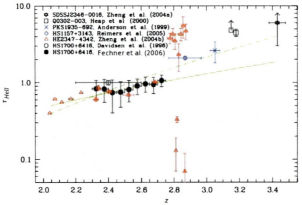

Fig. 8.2 HeII Gunn–Peterson optical depth as a function of redshift [81]. The solid line represents a fit to the data for $z < 2.75$, while the dashed line is the overall fit, which is characterized by a much steeper slope. The data from QSO HE 2347-4342 [311] show a significant increase in optical depth at $z \simeq 2.8$ (with permission from AA, Copyright (2006) ESO).

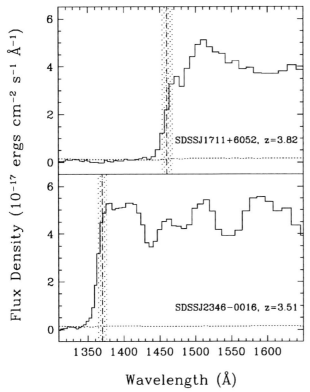

Fig. 8.3 UV prims spectra obtained with HST/ACS of quasars SDSS J2336-0016 at $z \simeq 3.51$ (lower panel) and SDSS J1711+6052 at $z \simeq 3.82$ (upper panel) [312]. The dashed lines mark the expected wavelength of He II Lyman α with the shaded region indicating the uncertainty due to the spectral resolution of the ACS UV prism. Both objects show quite dark troughs and SDSS 1711+6052 also displays what happears to be a damping wing (Reproduced by permission of the AAS).

8.3
Constraints from the Temperature of the IGM

We have seen in Section 4.4.1 that the core of the Lyman α cross section has a width given by the convolution of the intrinsic with the typical velocity dispersion of the gas (see also Figure 4.8). Thus, it is possible to measure the IGM temperature by studying the width of unsaturated Lyman α absorption systems. Reionization heats up the gas, so by measuring the IGM temperature as a function of redshift one can constrain the epoch of reionization [272]. This is shown in Figure 8.4. In general terms we should expect IGM heating from helium reionization to be less than that due to hydrogen reionization for two main reasons. One is that the ionization threshold is higher by a factor of four, so that relatively fewer photons are much more energetic than the threshold and this will lead to a lower excess energy. The second reason is that helium has a small-

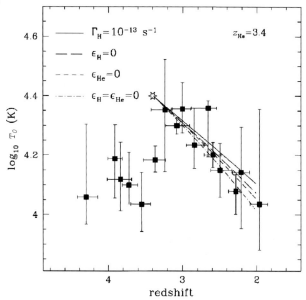

Fig. 8.4 IGM temperature data compared to HeII reionization models. All models assume reionization at $z = 3.4$ and the starting point adopted is indicated by a star. The solid line is a model with photoheating from HI, HeI, and HeII. The short dashed curve reflects heating from HeII. The short dashed neglects HeII, while the long dashed neglects both HI and HeI. The dot-dashed curve ignores photoheating [272] (Reproduced by permission of the AAS).

er number density than hydrogen. For a Population III star with $T = 10^5$ K we find the mean (unweighted) excess energy to be 11.06 eV compared to the value of 15.95 eV for the case of hydrogen (see Table 3.1). This is less of an issue for QSOs that have a mean excess energy of 22.8 eV. Considering the number density of helium these values reduce to 0.96 eV or 1.98 eV per particle, respectively, for Population III stars and QSOs. Despite the decreased heat input compared to hydrogen, helium reionization occurs at a redshift where inverse Compton cooling is no longer effective so that the IGM can remain warm for a longer time.

8.4
Change in Metal-Line Ratios

It has been observed that the ratio of SiIV/CIV column densities changes at $z \approx$ 3–3.5 [249,250] despite the CIV column densities remaining roughly constant from $z = 2$ to $z = 5$. This suggest that the ionizing spectrum is changing and the most natural reason is that the reionization of HeII makes the IGM transparent to photons more energetic than 54.4 eV. This change in the spectrum alters the line ratios and can be used to constrain the reionization of helium.

8.5
Change in HI Lyman α Forest

Analysis of a thousand QSO spectra from the SDSS reveals that the effective optical depth shows a dip at $z \sim 3$ in what is otherwise a smooth behavior [22]. This is shown in Figure 8.5. This result would be quite intriguing if confirmed, as at the same redshift the temperature of the IGM is also seen to increase and both these effects could be explained by the reionization of HeII.

8.6
Reionizing Hydrogen First and Helium Later

In order to reionize and keep ionized hydrogen at redshift $z \lesssim 6$ without fully ionizing HeII one needs a soft UV spectrum. The issue here is whether one can observe a Gunn–Peterson trough for helium while one for hydrogen is not present. One way to establish whether this is possible is to compare the two optical depths. Madau and Meiksin [151] found that the connection between the Gunn–Peterson

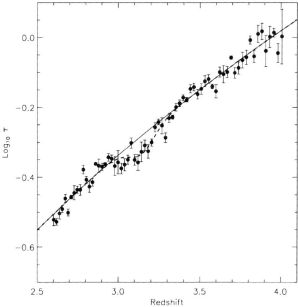

Fig. 8.5 The variation of the effective optical depth (including the effect of the Lyman α forest) as a function of redshift. The points show binned QSO measurements. The solid line is a smooth interpolation, while the dashed line is a fit to the data [22]. There is a dip at $z \sim 3$ that can be interpreted as a consequence of the reionization of HeII (Reproduced by permission of the AAS).

optical depth for HeII and HI is:

$$\tau_{GP}^{HeII} \approx 0.45 S_L \tau_{GP}^{HI} \tag{8.1}$$

where the spectral softness S_L can be defined as the flux ratio between 912 Å and 228 Å. QSOs and Population III stars have $S_L \gtrsim 2$, which would imply similar Gunn–Peterson optical depths for hydrogen and helium II. By comparison, stars at $T_{eff} \simeq 5 \times 10^4$ K have $S_L \simeq 212$. The value of S_L becomes much larger at lower temperatures. Thus, complete ionization of hydrogen by a hard spectrum like that of AGNs or Population III stars or, more importantly, a hard UV background keeping hydrogen ionized would be problematic as it would also ionize and keep ionized helium. A hard ionizing background would perhaps be possible if somehow, after reionization, the UV background spectrum became softer so that it could keep hydrogen ionized while allowing helium II to recombine at least in part. Note that since HeII is hydrogenoid, there is a simple rescaling for the effective recombination coefficients [197] namely:

$$\alpha_{He}(T) = Z_N \alpha_H(T/Z_N^2) \tag{8.2}$$

where $Z_N = 2$ is the charge of the helium nucleus, and T is the temperature. Using (2.11) and (8.2) we find that the effective recombination coefficient for helium is, to within 10%, ~ 6 times larger than that of hydrogen for temperatures between 5×10^3 K and 3×10^4 K. On this basis, it is conceivable that helium could recombine faster than hydrogen after having being initially reionized. Thus, the door is not completely closed on the possibility of a major contribution of QSOs to reionization of hydrogen as long as the ionizing output of stars increases fast enough to become the dominant contribution to the UV background below $z = 6$. A problem with this scenario is that the luminosity density associated to QSOs appears to increase with decreasing redshift between $z = 5$ and $z = 3$ at least as much as that of galaxies, so that it is unclear how QSOs could contribute to reionization but not be a major contributor to the UV background after reionization.

8.7
A Limit on the Escape Fraction from Galaxies at $z \simeq 3$

By following the arguments of the previous section and by studying the properties of the Lyman α forest at $z \lesssim 4$, it is possible to place limits on the escape fraction of galaxies. This is done by balancing the required UV background with that generated by known QSOs, adopting the observed UV luminosity density of Lyman-break galaxies and reasonable SEDs, and assuming values for the escape fraction [25]. The resulting constraints are shown in Figure 8.6. This type of models support a mean galaxy escape fraction $f_c \lesssim 0.1$ and consequently a relatively hard UV background at $z \simeq 3$. This provides an argument in favor of hydrogen reionization by stars as the galaxy contribution is more significant than that of AGNs at $z \gtrsim 4$.

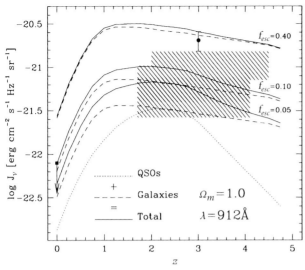

Fig. 8.6 UV background for models with galaxies escape fraction $f_c = 0.05, 0.1, 0.4$ (dashed lines). The QSO contribution is given by the dotted line. The resulting total background is given by the solid lines. The shaded area represents the Lyman limit UV background estimated from the QSO proximity effect [59, 96, 234]. The arrow is an upper limit to the local ionizing background [288]. The data point at $z = 3$ is from Lyman-break galaxies [258]. More recent masurements of LBGs suggest lower escape fractions. Values of escape fraction $f_c \lesssim 0.1$ are in better agreement with the data [25] (with permission from AA, Copyright (2001) ESO).

9
Future Instrumentation

9.1
Overview

A number of new astronomical facilities under development or study are capable of making important contributions to the investigation of first light and reionization. In the area of traditional visible-near-IR astronomy we will review the James Webb Space Telescope, the Wide-Field Camera 3 to be installed on the Hubble Space Telescope and the potential capabilities of a future large space telescope. We will also discuss the capabilities of the future large ground-based telescopes. In the area of radio-astronomy we will consider the MWA, LOFAR, ALMA, and a possible moon-based radio telescope. Finally, we will consider large field of view telescopes from the ground and from space.

9.2
The James Webb Space Telescope

The James Webb Space Telescope (JWST) is a large, passively cooled, cryogenic, space telescope designed to operate at the Lagrangian point L2 of the Earth–Sun system, at about 1.5×10^6 km from the Earth. A large sunshield prevents the cold optics of the telescope from being illuminated by the Sun, the Earth, or the Moon. JWST has a segmented primary mirror with a collecting area of 25 m^2. Figure 9.1 shows the current configuration of JWST. The telescope passed its NASA confirmation review in mid-2008, entering the nominal construction phase. JWST is built by NASA in partnership with the European Space Agency (ESA) and the Canadian Space Agency (CSA). The Space Telescope Science Institute that currently operates HST will be the Science Operations Center also for JWST. Here, we will provide short historical highlights on the project, and a description of the telescope, its capabilities and its science goals. More information can be found on the NASA and STScI web sites and in the JWST primer maintained by STScI [262].

From First Light to Reionization. Massimo Stiavelli
Copyright © 2009 WILEY-VCH Verlag GmbH & Co. KGaA, Weinheim
ISBN: 978-3-527-40705-7

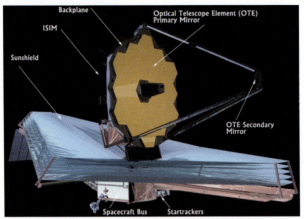

Fig. 9.1 The James Webb Space Telescope in its design configuration at mission confirmation. The segmented mirror in gold-coated beryllium is visible in the center. The multilayer sunshield separates the cold optics from the warm instrument module and the spacecraft bus visible below the sunshield. The instrument module (ISIM) is directly attached to the backplane of the primary mirror (courtesy of NASA).

9.2.1
Historical Remarks

A first conference on a successor of HST, dubbed at the time the Next Generation Space Telescope (NGST) was held at the STScI in Baltimore in 1989 [20]. After a hiatus of a few years, the release of the HDFN highlighted the potential of space telescopes in studying galaxies at high redshift. The next logical step was improved capabilities in the infrared. The concept of a large infrared telescope in space was one of the recommendations of the *HST and Beyond* committee, chaired by Alan Dressler [74]. Following the release of this report, a detailed study phase started in earnest and led to the development of a further report capturing the science ideas and design concepts for the telescope [268]. These and other reports were provided as input to the National Academies Decadal study providing prioritization for space- and ground-based astronomy projects [166]. The commitee placed NGST as the top priority among astronomical space missions and NASA officially started the study phase for the project. In the early phases of the project, science input was provided first by an ad hoc Science Working Group and later by an Interim Science Working Group until the flight Science Working Group could be formed in 2002 at the same time that Northrop Grumman Space Technology was selected as the prime contractor and the project was renamed the James Webb Space Telescope after the NASA Administrator in the Apollo era.

The original mirror diameter recommended for NGST was 4 m [74] but it grew to 8 m during the following studies [268] leading to the National Academy endorsement. Following the beginning of the NASA design phase, the diameter shrunk to the present 6.5 m equivalent.

9.2.2
The JWST Science Requirements Document

The design of JWST has been driven by requirements derived by considering four main science themes:
- First light and reionization. JWST will study the reionization history of hydrogen and identify the sources (galaxies or AGNs) responsible for reionization. JWST will also identify the first galaxies to form in the Universe and study their properties.
- Assembly of galaxies. JWST will study how galaxies form and evolve and are assembled in the morphological sequence we see today. JWST will also study the interplay between galaxy evolution and nuclear activity and the origin of the global correlations of galaxy properties.
- Formation of stars and protoplanetary disks. JWST will study the protostellar collapse, the effects of environment on galaxy formation, and the origin of the initial mass function. JWST will also study how gas and dust coalesce to form protoplanetary disks.
- Planets in the solar system and beyond. JWST will study how planets form and their physical and chemical properties. JWST will investigate how habitable zones are established and study extrasolar planets through coronagraphy and transit observations.

In the following we will focus mostly on the first theme which is the most relevant to the subject of this book. Within this science area, JWST is designed to answer a few high-level questions such as:

1. When did the first luminous sources arise and what was their nature? What where their properties?

2. When did reionization occur? What is the reionization history of the Universe prior to final reionization?

3. What were the sources responsible for reioniziation? Were they powered by nuclear fusion or gravitational accretion?

The answer to question 1 requires deep imaging to identify galaxies at the highest possible redshift and followup observations to study their properties. The answer to question 2 involves obtaining $R \simeq 3000$ spectra of QSOs that will be discovered at $z \gtrsim 7$ and the study of the evolution of line ratios such as Lyman $\alpha/H\alpha$. Finally, question 3 requires estimating the hardness of the ionizing radiation of galaxies at high z. To carry out these measurements JWST requires a combination of near-infrared and mid-infrared imaging and spectroscopy. The highest-redshift galaxies seen in the HUDF are barely resolved and very faint. On this basis, JWST is designed for very high sensitivity but an angular resolution in the near-infrared comparable to what HST achieves in the visible.

9.2.3
Overview of JWST Instrumentation

JWST is designed to take images and spectra at red–visible wavelengths and in the near- and mid-infrared. The integrated science instrument module (ISIM) includes a near-IR camera (NIRCam), and near-IR spectrograph (NIRSpec), a mid-IR instrument capable of taking both images and spectra (MIRI) and finally the tunable filter instrument (TFI) within the fine guidance sensor (FGS). The TFI is capable of obtaining $R = 100$ narrow-band images at any wavelength selected by the user.

NIRCam. The NIRCam instrument is built by Lockheed ATC under the leadership of Marcia Rieke at the University of Arizona. NIRCam is made of two symmetric modules each containing a 'blue' and a 'red' wavelength channel taking images simultaneously using a dichroic. The blue channel covers wavelengths shorter than 2.5 μm, while the red channel covers from 2.5 to 5 μm. The field of view of each module is 2.2×2.2 arcmin so that the instantaneous field of view of the camera is ~9.6 arcmin^2 but the effective field of view for a wide area multiwavelength survey is ~19 arcmin^2 thanks to the ability of obtaining simultaneous images in two bands. All channels use HgCdTe arrays with a long-wavelength cutoff of 2.5 μm for the blue channel and 5 μm for the red channel. The short-wavelength cutoff is set by the gold coatings on the JWST optics.

JWST and NIRCam do not have enough sensitivity to detect individual Population III stars that are expected to have AB magnitudes of ~39. The focus then will be on the first galaxies or, possibly, on the first star clusters. The field of view of NIRCam is large enough that it should contain ~10 galaxies at $z \approx 10$ [279] at AB magnitudes around 31 or possibly brighter. Unfortunately, the same models predict that the surface density of galaxies at higher redshift decreases rapidly so that NIRCam might find them only if their location is known a priori from the location of, e.g., a gamma-ray burst. The existing models are quite uncertain at high z but – as we have seen in Chapter 6 – a rapid evolution of the luminosity function of Lyman-break galaxies at $z \gtrsim 6$ appears to be supported by observations [33, 186].

NIRSpec. The NIRSpec instrument is built by Astrium under contact of the European Space Agency and the leadership of Peter Jakobsen. NIRSpec is a multiobject spectrograph using as programmable slits an array of microshutters and is capable of obtaining simultaneous spectra of 100 or more objects at resolving powers of about 100, 1000, or 2700. The field of view of the instrument is about 10 arcmin^2. The detectors are the same as the red channel of NIRCam with a long-wavelength cutoff of 5 μm. As we have seen in Figure 7.12 the majority of the galaxies imaged today by HST in the HUDF are beyond the capability of spectroscopic followup from the ground. In very long integrations of about 10^6 s, NIRSpec should be able to reach ABmagnitudes of ~29 and therefore study all these objects. The fainter galaxies down to ABmagnitude ~31 that will be imaged in the JWST ultradeep surveys are beyond the capabilities of NIRSpec, but perhaps they could be studied by lensing amplification.

MIRI. MIRI is a mid-IR instrument built jointly by JPL and a European Consortium under the joint leadership of George Rieke (University of Arizona) and Gillian Wright (UK Astronomy Technology Center, Edinburgh). Once a sample of very faint galaxies at very high z has been obtained using, e.g., the Lyman-break technique with NIRCam and once their redshift has been confirmed with NIRSpec, we will need to assess whether these objects are primordial, i.e. whether they are experiencing their first burst of star formation. If the objects are too faint to measure their metallicity with NIRSpec, one might resort to the indirect test of looking for the presence of an underlying older stellar population. Such a population would be most easily detected longwards of the 4000 Å rest frame. For redshifts greater than ~11 this requires going beyond 5 μm, in the wavelength range of the MIRI instrument. MIRI is capable of imaging and spectroscopy. Given the faintness of our expected targets imaging should be the most generally applicable capability of MIRI but, for a few sources, it will be interesting to study the Hα line to compare its strength to that of Lyman α. At redshifts greater than ~6.6 the study of the Hα will require the spectroscopic capabilities of MIRI.

TFI. The TFI is built by the Canadian Space Agency under the leadership of John Hutchings and Rene Doyon. We have seen in Chapter 5 that Lyman α sources are one of the tools to study reionization. One of the most powerful means to identify these sources is the narrow band excess technique using narrow-band filters. The same technique can be used by JWST using the Tunable Filter Instrument that is capable of $R \sim 100$ imaging at user selectable wavelength longer than about 1.5 μm. The detectors are the same as the long-wavelength cutoff HgCdTe arrays used also in NIRCam and NIRSpec.

9.3
Other Space-Based Instrumentation

It is useful to consider what will be available before and after JWST. Before JWST is launched the HST's capabilities will be upgraded during Servicing Mission 4 with the installation of the WFC3 and COS instruments that we describe below. In the more distant future, the long development timescales of space missions require us to study a possible future large telescope even before JWST is launched.

9.3.1
The Wide-Field Planetary Camera 3

The Wide-Field Planetary Camera 3 (or WFC3) is an instrument for HST that should be already installed on Hubble during Servicing Mission 4 by the time this book is in print. WFC3 includes two imaging channels, the UVIS channel operating across the full sensitivity range of its two silicon CCDs from 2000 Å to 10000 Å and the IR channel using HgCdTe infrared arrays with a long-wavelength cutoff set

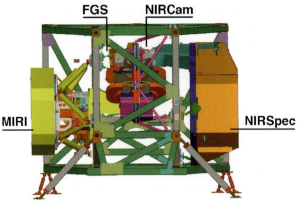

Fig. 9.2 A schematic drawing of the JWST ISIM module with the location of the JWST instruments (courtesy of NASA).

at ~1.7 µm. This was chosen because the HST primary and secondary mirrors are kept at 15–17 °C for thermal-stability reasons and generate a thermal background, so that HST is no longer competitive in the K band. The choice of a shorter cutoff also enables a simpler instrument design, such as using thermoelectric cooling instead of cryogenics. The UVIS channel has a field of view of 7.5 arcmin^2, while the IR channel has a field of view of 4.9 arcmin^2. For the type of high-redshift studies that are the focus of this book, the IR channel is the most relevant and is provided with a set of filters that should enable efficient Lyman-break selections of the type discussed in Chapter 6, as well as with two grisms to obtain slitless spectroscopic data.

9.3.2
The Cosmic Origins Spectrograph

The Cosmic Origins Spectrograph (COS) is a high-efficiency, high resolving power ($R \sim 20000$) UV spectrograph optimized for studying absorption line systems in QSOs. In the context of the study of reionization, the higher sensitivity of COS will enable us to study with greater detail the helium II forest, improving our constraints on the ionizing spectrum and therefore, indirectly, on the relative importance of galaxies and active galactic nuclei in establishing the UV background that maintains the IGM ionized.

9.3.3
A Possible Future Large Telescope in Space

JWST cannot detect individual Population III stars because they are too faint. Let us determine how large would a telescope need to be in order to detect them. The signal-to-noise ratio for a point source with count rate per unit area s observed by a telescope of diameter D in a background with a count rate per unit solid angle of b and with a detector with an effective noise rate per pixel of d is given by:

$$\frac{S}{N} = \frac{s\pi(D/2)^2 \sqrt{t_{exp}}}{\sqrt{(s + bN_p(\lambda/D)^2)\pi(D/2)^2 + dN_p}} \tag{9.1}$$

where N_p is the effective number of noise pixels and t_{exp} is the exposure time and λ is the wavelength of observation. The effective noise includes the effect of dark current and read-out noise. The latter is independent of time for a single image but becomes time dependent for long integrations with many images. Equation (9.1) accounts for the fact that the signal from the source and the background is proportional to the area of the telescope $\pi(D/2)^2$ but, for a Nyquist sample system, the pixel size has an angular size decreasing as D^{-1}, so that the sky area contributing to the background in each pixel goes as D^{-2}. In the limit of faint sources, the

Fig. 9.3 WF/PC camera seen here during the first Servicing Mission to HST in December 2003 as it is removed from HST to be brought back to the ground. After spending several years under nitrogen purge at JPL several components of this instruments were reused to construct the WFC3 instrument, which will fit the same radial bay on HST (courtesy of NASA).

contribution of s to the noise can be neglected as $s \ll bN_p(\lambda/D)^{-2}$. Also, in the background-limited case, $b < b\lambda^{-2} > d$ so that also the detector noise term can be neglected. Within this approximation the S/N can be written as:

$$\frac{S}{N} \simeq \frac{sD^2\sqrt{\pi t_{exp}}}{2\sqrt{bN_p\lambda^2}} \qquad (9.2)$$

By varying the telescope diameter while keeping the S/N constant, we can use (9.2) to show that for background-limited imaging of faint sources the exposure time varies as the fourth power of the telescope diameter. For constant exposure time and signal-to-noise ratio we find instead that the source count rate s decreases with D^2. JWST can observe sources at $AB \sim 31$, while Population III stars at $AB \sim 39$ are about 1600 times fainter. Thus, a space telescope able to observe individual Population III stars needs a diameter $\sim \sqrt{1600}$ times larger, or about 260 m. Placing the telescope on a 3 AU orbit instead of in L2 would decrease the background b by factor of the order of 3^4 where a factor 3^2 is due to a decrease of the sunlight at 3 AU, inversely proportional to the square of the distance, and the other factor 3^2 is due to the change in properties of the zodiacal disk. Thus, a telescope at 3AU able to detect individual Population III stars would have a diameter of about 87 m. Needless to say both a 260 m telescope at L2 or an 87-m one at 3 AU are well beyond present technical and budgetary capabilities. Detection of a cluster of 40 Population III stars (see Section 6.4) would *only* require sensitivity down to $AB \sim 35$, which could be achieved by a space telescope at L2 of 40 m in diameter or one at 3 AU with a 13-m primary mirror.

JWST can image galaxies down to $AB \sim 31$ but can take low-resolution ($R \sim 100$) spectra of galaxies 'only' down to $AB \sim 29$. Let us estimate how large a telescope would need to be to take spectra down to the imaging limit of JWST. For point sources and for resolving power R the S/N ratio equation is identical to (9.2), except that s and b are replaced by s/R and b/R, respectively. Thus, in order to gain 2 magnitudes in sensitivity for point sources, one needs a diameter $\sqrt{6.5}$ larger than JWST or about 16 m. For extended objects like galaxies, (9.2) needs to be modified as the area to be considered is no longer determined by the diffraction limit and the number of effective pixels, but is set by the size of the object so that:

$$\frac{S}{N} \simeq \frac{sD\sqrt{\pi t_{exp}}}{2\sqrt{bA}} \qquad (9.3)$$

where A is the solid angle subtended by the target galaxy. This expression is valid when the target galaxy is resolved, i.e. $A > (\lambda/D)^2$. Equation (9.3) shows that gaining sensitivity for extended objects is harder than for point sources and the same 2 magnitudes in sensitivity increase for extended galaxies would require a 40-m telescope.

The Advanced Technology Large-Aperture Space Telescope (ATLAST) is a 16-m UV-optical telescope concept that can be extended easily to the near-IR [213]. Such a telescope would have a variety of scientific applications and in the context of the study of first light and reionization would be a powerful followup to JWST. While still unable to detect individual Population III stars without extremely long integrations it would be able to obtain low-resolution spectra for all compact sources imaged by JWST.

9.4
Large Ground-Based Telescopes

For a given telescope diameter, it is much easier to build a telescope on the ground than in space. Below, we will describe the relative advantages and disadvantage of the two locations and a few potential capabilities of a future large ground-based telescope, especially those relevant to the subject of this book. We will focus mostly on imaging and low to medium resolution spectroscopy. Ground-based telescopes are generally preferred for high resolving power spectroscopy that is always detector limited.

9.4.1
Ground vs. Space Comparison

The space environment is more stable and characterized by lower backgrounds. Thus, imaging is more naturally done from space. Spectroscopy is more dependent on excellent detector properties and by dispersing the background is less sensitive to high backgroundlevels. Thus, a ground-based telescope is favored for high-resolution spectroscopy when this becomes detector limited. A ground location also makes it easier to swap detectors and benefit from improvements in detector technology that can be exploited in space only by launching a new instrument or, at least, carrying out a servicing mission for telescopes like HST, designed to be serviced on orbit.

Fig. 9.4 An artist's view of the TMT telescope. A person and a car are provided for comparison (courtesy of the TMT *observatory* corporation).

We saw that imaging a cluster of Population III stars requires a 40-m space telescope at L2. We can estimate how large a ground-based telescope would need to be in order to detect such a cluster. Let us assume that adaptive optics with high Strehl ratio is available. The Strehl ratio is the ratio of the peak value in the point-spread function divided by the peak value of the same quantity for an ideal telescope with the same diameter. At present, Strehl ratios achieved with adaptive optics are below the value of 0.8 customarily considered to define the diffraction limit and achieved, e.g., by HST imaging instruments. However, in the following we will disregard this difference and assume that adaptive optics technology will bridge this gap. The background b on the ground in the H band is about 800 times larger than from space. From (9.2), the resulting decrease in S/N by $\sqrt{800}$ needs to be compensated by an aperture ~5 times larger so that a ground-based telescope able to detect a cluster of Population III stars would need to about 200 m in diameter.

Let us consider instead the capability to image galaxies. Faint galaxies are barely resolved at the HST and JWST resolution and they would be over-resolved by a large ground-based telescope. Thus, we need to use (9.3) that takes into account the finite size of galaxies. Now, the limiting flux varies linearly with the telescope diameter instead of the square. In order to match JWST's imaging down to $AB \sim 31$ a ground-based telescope with its 800 times higher background would need to have a diameter $\sqrt{800}$ times larger than JWST, or about 180 m. Both of these imaging estimates would be dramatically different if one found a way to suppress the OH emission lines that provide a large fraction of the background from the ground. Unfortunately, so far a high-throughput, multiline, OH suppression system has not yet been demonstrated. Another potential difficulty with adaptive optics is that the PSF often shows wings at about the size of the natural seeing, i.e. at about 1 arcsec in diameter. Already in the HUDF, one galaxy is detected every 4 square arcsec so that in the deeper JWST fields we should expect even more crowded images. At this level of crowding, overlapping wings of the PSFs would become a serious issue unless the Strehl ratio can be kept very high.

Spectroscopy at a resolving power of $R = \lambda/(\Delta\lambda) \sim 100$ is background limited both from space and from the ground. Therefore, the same arguments made for imaging are also applicable for low-resolution spectroscopy. Let us consider instead spectroscopy at $R > 3000$. At this resolving power both a space telescope and a ground-based telescope operating in between the OH lines are detector limited. In this case the relevant equation for a point source is:

$$\frac{S}{N} = \frac{(s/R)\pi(D/2)^2\sqrt{t_{exp}}}{\sqrt{dN_p}} \tag{9.4}$$

With the background out of the picture only the detector noise properties (and the instrument throughput that we have ignored so far) play a role and a ground-based telescope would gain in S/N over JWST with the square of the ratio of the two diameters.

For spectroscopy of extended objects (9.5) needs to be changed to account for the fact that the diffraction-limited pixels would over-resolve the galaxy so that we have:

$$\frac{S}{N} = \frac{(s/R)\pi(D/2)^2 \sqrt{t_{exp}}}{\sqrt{d(A\,D^2/\lambda^2)}} \tag{9.5}$$

where the term AD^2/λ^2 captures the number of resolution elements covering the galaxy. Thus, for a resolved galaxy we are once again with a sensitivity simply linear with the diameter. A seeing-limited telescope not resolving the galaxy would place a smaller number of pixels over the object and would have higher sensitivity.

A lesson from the above considerations is that high resolving power spectroscopy is indeed best done from the ground at all wavelengths accessible from within the atmosphere. For both imaging and for obtaining spectra of galaxies, over-resolving these objects is counterproductive to sensitivity and leads to a flux limit decreasing as the telescope diameter instead of the diameter squared. Thus, there seems to be a strong science justification for seeing limited instrumentation even on the next generation of ground-based telescopes.

9.4.2
Multiobject Spectroscopy

Very high redshift galaxies are thought to be rare and faint. Obtaining their spectra will require long integration times and this would make it convenient to obtain simultaneously as many spectra as possible. The areal density of 'bright' i-dropout galaxies is about 1 per square arcmin and the density of $z = 10$ galaxies at the limit of a JWST UDF survey is similar. This suggests that the field of view of a spectrograph capable of obtaining simultaneous spectra of several of these objects should be at least ~10 square arcmin. For observing galaxies at $R > 3000$ there is no benefit from adaptive optics since the spectra are detector limited and using natural seeing would make it easier to obtain a large field of view by allowing larger pixel sizes.

9.4.3
Very High Resolution Imaging

For high surface brightness objects, adaptive optics from the ground can provide a much higher angular resolution than is available from a space telescope with a smaller aperture and this is an important capability to study detailed structures in galaxies. This may unfortunately be more important for $z \lesssim 5$ as at $z \gtrsim 6$ cosmological surface brightness dimming will make the resolved parts of galaxies very faint.

9.5
Observing 21-cm Radiation at High Redshift

We have seen in Chapter 5 that 21-cm radiation can be a powerful diagnostics tool for probing the dark ages. Several facilities to do so have been proposed or are under development and we will attempt below to review some of the major ones.

9.5.1
Murchison Wide-Field Array

The Murchison Wide-Field Array (MWA), formerly known as the Mileura Widefield Array is located in Western Australia in one of the sites with the lowest radio pollution on the planet and its goal is to operate between 80 and 300 MHz, enabling it to study the 21-cm line up to $z \simeq 16.8$. The MWA is a joint project between MIT, the Harvard-Smithsonian Center for Astrophysics, a consortium of Australian Universities and the Raman Research Institute in India. The MWA has 500 tiles with 16 dipoles each, spread over ~1.5 km and with an effective collecting area of 8000 m². The field of view is of the order of tens of degrees and the angular resolution is 3.4 arcmin with a point-source sensitivity of 0.33 mJy at 250 MHz in one hour. It is expected that the MWA, after two years of data taking, should be able to constrain the ionized bubbles filling factor at an epoch when hydrogen is 50 per cent ionized [138].

9.5.2
Low-Frequency Array

The low-frequency array (LOFAR) is an international project led by the Netherlands in partnership with several European countries. LOFAR is an array of 7700 antennas covering a low band (30–80 MHz, see Figure 9.5) and a high band (110–240 MHz). The low band is sensitive to 21 cm in the redshift range $z \simeq 16.8$–46, while the high band is sensitive in the redshift range $z \simeq 4.9$–13. LOFAR has a core

Fig. 9.5 A testsite of the LOFAR core configuration showing a number of dipole tiles for the LOFAR low band (courtesy of the LOFAR project).

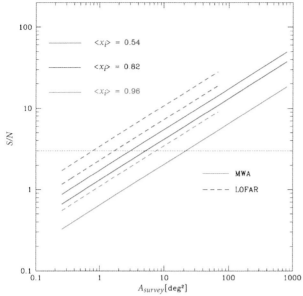

Fig. 9.6 Signal-to-noise ratio for the detection of the 21-cm galaxy cross-spectrum for different values of the ionization fraction $\langle x_i \rangle$. The solid line refers to MWA, while the dashed lines refer to LOFAR [139]. The higher sensitivity of LOFAR enables it to achieve significant detections even for surveys area of ten square degrees (Reproduced by permission of AAS).

configuration with 3200 antennas and an extended array of 4500 antennas on much larger baselines. The expect sensitivity in one hour of integration of the full array goes from 2 mJy at 30 MHz to 0.06 mJy at 200 MHz. LOFAR, when combined with a Lyman α search carried out at, e.g., Subaru, is capable of detecting the 21-cm galaxy cross-spectrum as shown in Figure 9.6.

9.5.3
Square-Kilometer Array

The square-kilometer array (SKA) is a new radiotelescope concept being studied as an international collaboration including the United States, the United Kingdom, Germany, Italy, the Netherlands, Poland, Sweden, Australia, Canada, China and India. SKA will be operating from 100 MHz to 25 GHz or an even wider range with a large field of view and high sensitivity. The low-frequency limits of SKA will limit it to lower redshifts than those accessible to, e.g., LOFAR's low band. However, where SKA is sensitive, it is more powerful than either MWA or LOFAR [46].

9.5.4
A Radiotelescope on the Far Side of the Moon

Observation of the 21-cm absorption over the CMB at redshifts in the range $z \simeq$ 50–200 requires being able to observe at frequencies lower than 30 MHz with very high sensitivity. Effective observation below 30 MHz requires going to space and an orbiting radiotelescope would probably be viable. However, an observatory near one of the Moon poles but on the far side of the Moon would have the added advantage of being shielded from the Earth by several kilometers of rock, cutting down dramatically terrestrial radio interference. Clearly the costs of such an observatory would be significant but so would be its discovery potential.

9.6
Atacama Large Millimiter Array

The Atacama Large Millimiter Array (ALMA) is a facility developed in partnership by several countries and organizations including the United States, Canada, Japan, Spain, and the European Southern Observatory. ALMA is sited at 5000 m altitude near Cerro Chajnator in Chile. The high sensitivity and high angular resolution of ALMA will enable it to study the continuum emission of galaxies especially if these objects contain dust. One generally expects high-z objects to be metal poor and therefore also relatively dust free but this statement is valid only on average. Specific objects could have evolved faster than the average and self-enriched even at high redshift. Indeed, some of the SDSS QSOs at $z \gtrsim 6$ have high metallicity [89] and dust [24]. ALMA will be able to study these objects in great detail. Still, detecting a first-light galaxy candidate with ALMA would cast serious doubts on its real nature. ALMA will also be able to measure the redshifts of star-forming galaxies using CO emissions and this might become an efficient way to obtain galaxy redshifts.

9.7
Large Field of View Imaging

Some of the objects of interest for the study of the dark ages are rare but relatively bright. Examples are high-z QSOs and pair-instability supernovae. Once discovered, facilities like JWST will be able to study them in great detail. However, the chances that JWST will find them are very slim given their rarity. In order to find these objects, we will have to rely on existing large field of view instruments, especially those on 8-m class telescopes, or on the next generation of large field of view instrumentation such as the LSST or the dark energy missions JDEM (NASA) or Euclid (ESA) especially if they will include near-IR widefield imaging capabilities.

9.7.1
Large Synoptic Survey Telescope

The Large Synoptic Survey Telescope (LSST) is an optical telescope with an effective aperture of 6.8 m designed for large field of view imaging observations. Its field of view is 9.6 square degrees. LSST will be built in Cerro Pachón in Chile and should be able to image 20 000 square degrees of sky about 100 times per year, obtaining variability data for a wide variety of astronomical sources. Using red-sensitive CCD detectors these data should enable us to identify a large number of $z \simeq 7$ QSOs. Low-redshift pair-instability supernovae from late Population III stars, forming from primordial material surviving uncontaminated in voids, would also be detectable [275]. The estimated rate of high-redshift pair-instability supernovae is about 1 per year per 400 square degrees (see Section 6.8). If Population III stars continue to form at redshift $z \simeq 6$–7 with similar rates, the LSST would be able to detect such supernovae despite their slowly varying light curve.

9.7.2
Large Field of View Imaging from Space

Large field of view imagers from space have been proposed as possible dark-energy missions based on their capability to identify large numbers of supernovae and to study weak lensing. An an example, the SuperNova Acceleration Probe (SNAP) – a dark-energy mission study headed by S. Perlmutter – would have a field of view of 1 square degree. This capability, especially if available also in the near-IR, would be very powerful to identify pair-instability supernovae also at $z \gtrsim 7$, which is beyond the reach of silicon CCD detectors. The benefit of space in the near-infrared would once again be the low background, as reaching $AB \sim 27$ over hundreds of square degrees, as needed to identify the supernovae, would be challenging from the ground.

9.8
Planck

Planck is a European Space Agency mission designed to image the anisotropies of the cosmic microwave background over the whole sky with superior sensitivity and angular resolution. Launch is planned in the Spring of 2009. Planck will have broad impact on cosmology, will refine our cosmological parameters, and will identify many clusters of galaxies. Planck will also be able to refine the measurement of the Thompson optical depth carried out by WMAP, improving the integrated constraint on reionization [211]. However, it is unclear how conclusive this measurement will be considering that integrated constraints have limited discriminating power [179].

A
Overview of Physical Concepts

In this appendix we give some useful cosmological formulae and introduce three physical concepts and equations used in the main text, namely, Saha's equation for the ionization equilibrium, the Lane–Emden equation for polytropic stars, and the two-fluid Jeans instability.

A.1
Cosmological Quantities

The equation for the Hubble parameter in the matter-dominated era [206] is:

$$H(z) \equiv \frac{\dot{a}}{a} = H_0[\Omega_m(1+z)^3 + \Omega_R(1+z)^2 + \Omega_\Lambda]^{1/2} \tag{A.1}$$

where $\Omega_R = 1/(H_0 a_0 R)^2$ is the curvature term, a is the scalelength of the Universe, a_0 its present value, and $a_0 R$ is the spatial radius of curvature; H_0 is the present value of Hubble's constant, $\Omega_m = 8\pi G \varrho_0/(3H_0^2)$ with ϱ_0 the present matter density in the Universe and $\Omega_\Lambda = \Lambda/(3H_0^2)$, where Λ is the cosmological constant as it appears in Einstein's equations. In the following, we will consider only a flat Universe for which $\Omega_R = 0$. For convenience we define the quantity $\dot{a}/a$ as $E(z)$:

$$E(z) = [\Omega_m(1+z)^3 + \Omega_R(1+z)^2 + \Omega_\Lambda]^{1/2} \tag{A.2}$$

so that $H(z)$ can be written as $H_0 E(z)$.

The redshift z is defined as $1 + z \equiv a_0/a(t)$ and the lookback time is computed as follows:

$$t(z) = \int_0^a \frac{da}{H_0 \dot{a}} = \int_z^\infty \frac{dz}{H_0(1+z)E(z)} \tag{A.3}$$

The angular size distance for a flat Universe is:

$$D_A(z) = cH_0^{-1} \int_0^z \frac{dz'}{E(z')} \tag{A.4}$$

The proper element of distance as a function of redshift is:

$$\frac{dl}{dz} = \frac{1}{(1+z)H(z)} \tag{A.5}$$

From First Light to Reionization. Massimo Stiavelli
Copyright © 2009 WILEY-VCH Verlag GmbH & Co. KGaA, Weinheim
ISBN: 978-3-527-40705-7

As an example, the Thompson optical depth of (2.13) is obtained as an integral of $\sigma_T n_e$ weighted by dl/dz.

The bolometric distance modulus is given by:

$$DM(z) = 25 + 5\log[3000(1+z)D_A(z)] - 5\log h \tag{A.6}$$

when computing the magnitude of an object in a given band the relation between apparent and absolute magnitude is:

$$m_{\nu(1+z)} = M_\nu + DM(z) - 2.5\log(1+z) \tag{A.7}$$

A.2
Saha Equation

We are interested in deriving the equilibrium condition for a reaction of the type:

$$A + B = C + \gamma \tag{A.8}$$

One possible example of a reaction of this type is the ionization–recombination equilibrium for hydrogen. From the point of view of thermodynamics, equality of the chemical potentials gives the equilibrium condition so that:

$$\mu_A + \mu_B = \mu_C + \mu_\gamma \tag{A.9}$$

Let us start out determining the chemical potential for thermal radiation. The energy of radiation in a volume V is:

$$E_\gamma = Va\,T^4 \tag{A.10}$$

where $a = 8\pi^5 k^4/(15h^3 c^3)$. In general, in thermodynamics the energy is a function $E(S, V, N)$ of the entropy S, of the volume V and of the number of particles N. The energy-related function that depends on the temperature is the free energy $F(T, V, N)$ that is obtained from E through a Legendre transform, namely:

$$F_\gamma = E_\gamma - TS_\gamma = E + T\frac{\partial F}{\partial T} \tag{A.11}$$

Equation (A.11) is a differential equation for $F(T, V, N)$ that has the solution:

$$F_\gamma = \frac{a}{3}T^4 V \tag{A.12}$$

The free energy F_γ does not depend on the number of particles N, so that the chemical potential is:

$$\mu_\gamma = \frac{\partial F}{\partial N} = 0 \tag{A.13}$$

which shows that – as expected – the chemical potential of thermal radiation is zero.

A.2 Saha Equation

Let us now compute the chemical potential for the other species. We start by writing a partition function Z_X for a simple particle system X with one energy level:

$$Z_X = g_X \cdot \exp[-\beta E_X] \int \exp\left[-\beta \frac{p^2}{kT}\right] d^3p \tag{A.14}$$

where the first term is the internal energy term and the second the kinetic term, g_X is the multiplicity of the one level with energy E_X and $\beta = 1/(kT)$. Integrating the first term we find the single-particle partition function:

$$Z_X = g_X \frac{(2\pi m_X kT)^{3/2}}{h^3} \exp\left(-\frac{E_X}{kT}\right) \tag{A.15}$$

where m_X is the mass of particle species X. The partition function for a system of N indistinguishable particles is $Z_{N,X} = Z_X^N/N!$ The connection between the free energy and the partition function is:

$$F_X = -kT \log Z_{N,X} \tag{A.16}$$

so the chemical potential for the species X is:

$$\mu_X = \frac{\partial F_X}{\partial N_X} = -kT \frac{\partial \log Z_{N_X,X}}{\partial N_X} \tag{A.17}$$

Using Stirling's approximation we can write $\log Z_{N,X} \simeq N_X \log Z_X - N_X \log N_X$ so that (A.17) becomes:

$$\mu_X = -kT \log Z_X/N_X \tag{A.18}$$

Using this equality and the fact that $\mu_\gamma = 0$, we can replace (A.18) into (A.9), divide by $-kT$, and take the exponential of both sides to find:

$$\frac{Z_A}{N_A} \frac{Z_B}{N_B} = \frac{Z_C}{N_C} \tag{A.19}$$

Replacing now (A.15) into (A.19) we find the equilibrium equation known as Saha's equation:

$$\frac{N_A N_B}{N_C} = \frac{g_A g_B}{g_C} \cdot \frac{m_A^{3/2} m_B^{3/2}}{m_C^{3/2}} \cdot \frac{2\pi kT}{h^3} \exp\left[-\frac{\Delta E}{kT}\right] \tag{A.20}$$

where $\Delta E = E_A + E_B - E_C$. Equation (A.20) reduced to (2.1) when species A is protons, B is electrons and C is neutral hydrogen atoms when we ignore the mass difference between protons and hydrogen atoms so that only the electron mass remains in the mass term.

A.3
Polytropic Stars

Let us derive the equation describing the hydrostatic equilibrium of a star governed by a polytropic equation of state $P = K\varrho^{1+1/n}$ connecting the pressure P to the density ϱ. K is a constant and n is the polytropic index. The case $n = 3$ corresponds to a relativistic gas, $n = 2/3$ corresponds to a nonrelativistic perfect gas.

The equilibrium equations for the spherically symmetric polytropic star are the Poisson equation:

$$\frac{1}{r^2}\frac{d}{dr}\left(r^2 \frac{d}{dr}\phi\right) = 4\pi G\varrho \tag{A.21}$$

and the equation of hydrostatic equilibrium that can be obtained by zeroing down time derivatives and velocity terms in Euler's equation (3.62) and writing the gradients in spherical coordinates ignoring the angular terms:

$$\frac{1}{\varrho}\frac{d}{dr}P + \frac{d}{dr}\phi = 0 \tag{A.22}$$

Following Chandrasekhar [51] we can pose $\varrho = \lambda\theta^n$, where λ is a constant and θ is a function of radius. With this definition the pressure becomes $P = K\lambda^{1+1/n}\theta^{n+1}$.

We can now eliminate the gradient of the potential ϕ in (A.21) by expressing it in terms of the pressure from (A.22). After some simple algebra, this gives the Lane–Emden equation:

$$\frac{1}{\xi^2}\frac{d}{d\xi}\left(\xi^2 \frac{d}{d\xi}\theta\right) = -\theta^n \tag{A.23}$$

having defined $r = a\,\xi$ with:

$$a = \sqrt{\frac{(n+1)K}{4\pi G}\lambda^{1/n-1}} \tag{A.24}$$

The Lane–Emden equation is solved using the boundary conditions: $\theta(0) = 1$ and $\frac{d\theta}{d\xi} = 0$ for $\xi = 0$. The solution to this equation is meaningful for $\theta \geq 0$. As the function θ is 1 for $\xi = 0$, there may be a radius where it becomes zero. The smallest radius where $\theta = 0$ is not known a priori and thus the Lane–Emden equation with its conditions at the center represents a free-boundary problem. For $n < 5$ the radius of the configuration is finite. Some analytical results can be obtained for $n = 0, 1, 5$ and $n > 3$. For generic n the equation has to be integrated numerically but this can be done simply using a Runge–Kutta integrator [215, 216].

Numerical integration for $n = 3$ gives a dimensionless radius $\xi_t \simeq 6.89685$, a ratio of central density to mean density $\varrho_0/\langle\varrho\rangle \simeq 54.1825$ and a central pressure:

$$P_c \simeq 11.05066\frac{GM^2}{R^4} \tag{A.25}$$

as a function of the physical mass M and radius R.

A.4
Jeans Instability for a Two-Fluid System

Let us consider here the case of two fluids, B for baryons and D for dark matter, interacting only by gravity and derive the stability conditions. We should note that dark matter is collisionless and it is best described by Vlasov's equation rather than fluid equations. However, the fluid equations can be recovered as moments of Vlasov's equations and therefore must also be valid. Moreover, they are much easier to deal with. In addition to fluid instabilities, there are other instabilities – like Landau damping – that cannot be studied in the fluid approximations.

The relevant equations are Poisson's equation for the gravitational field:

$$\nabla^2 \Phi = 4\pi G (\varrho_B + \varrho_D) \tag{A.26}$$

the continuity equations for the two fluids:

$$\frac{\partial \varrho_B}{\partial t} + \nabla \cdot (\varrho_B \mathbf{v}_B) = 0 \tag{A.27}$$

$$\frac{\partial \varrho_D}{\partial t} + \nabla \cdot (\varrho_D \mathbf{v}_D) = 0 \tag{A.28}$$

and Euler's equations for the two fluids:

$$\varrho_B \frac{\partial \mathbf{v}_B}{\partial t} + \varrho_B (\mathbf{v}_B \cdot \nabla) \mathbf{v}_B = -\nabla P_B - \varrho_B \nabla \Phi \tag{A.29}$$

$$\varrho_D \frac{\partial \mathbf{v}_D}{\partial t} + \varrho_D (\mathbf{v}_D \cdot \nabla) \mathbf{v}_D = -\nabla P_D - \varrho_D \nabla \Phi \tag{A.30}$$

Let us now perturb the system by expanding to 1st order (A.26)–(A.30) assuming that the unperturbed system is in equilibrium without bulk motions, i.e. $\mathbf{v}_B = \mathbf{v}_D = 0$ and that the perturbation has a spatial and time dependence of the type $\exp(i\mathbf{k} \cdot \mathbf{r} - i\omega t)$. Unperturbed quantities will be denoted with the $_0$ suffix, while the perturbed ones will be denoted by $_1$. The resulting equations are:

$$-k^2 \phi = 4\pi G (\varrho_{B,1} + \varrho_{D,1}) \tag{A.31}$$

$$-i\omega \varrho_{B,1} + \varrho_{B,0} i\mathbf{k} \cdot \mathbf{v}_{B,1} = 0 \tag{A.32}$$

$$-i\omega \varrho_{D,1} + \varrho_{D,0} i\mathbf{k} \cdot \mathbf{v}_{D,1} = 0 \tag{A.33}$$

$$-i\omega \varrho_{B,0} \mathbf{v}_{B,1} = -i\mathbf{k} P_{B,1} - \varrho_{B,0} (i\mathbf{k}\phi_1) \tag{A.34}$$

$$-i\omega \varrho_{D,0} \mathbf{v}_{D,1} = -i\mathbf{k} P_{D,1} - \varrho_{D,0} (i\mathbf{k}\phi_1) \tag{A.35}$$

We can now eliminate $\mathbf{v}_{B,1}$ and $\mathbf{v}_{D,1}$ by substitution, multiplying the linearized Euler equations by $\mathbf{k}$ and replacing the $\mathbf{k} \cdot \mathbf{v}_1$ terms using the continuity equations. The variable ϕ can be eliminated from Euler's equations using (A.31) and finally we can eliminate the pressure terms by defining the sound speed $c_B^2 = dP_B/d\varrho_B$ and $c_D^2 = dP_D/d\varrho_D$. These manipulations give:

$$k^2 c_B^2 \varrho_{B,1} - \omega^2 \varrho_{B,1} = 4\pi G \varrho_{B,0} (\varrho_{B,1} + \varrho_{D,1}) \tag{A.36}$$

and

$$k^2 c_D^2 \varrho_{D,1} - \omega^2 \varrho_{D,1} = 4\pi G \varrho_{D,0}(\varrho_{B,1} + \varrho_{D,1}) \quad (A.37)$$

We can now divide (A.36) by $\varrho_{B,1}$ and divide (A.37) by (A.36) to derive an expression for $\varrho_{D,1}/\varrho_{B,1}$. Equation (A.36) so modified becomes:

$$k^2 c_B^2 - \omega^2 = 4\pi G \varrho_{B,0} \left(1 + \frac{\varrho_{D,0}}{\varrho_{B,0}} \frac{k^2 c_B^2 - \omega^2}{k^2 c_D^2 - \omega^2}\right) \quad (A.38)$$

which gives us the dispersion relation. For $\varrho_{D,0} = 0$ we recover the familiar Jeans instability condition for a simple fluid. Equation (A.38) can be rewritten as a fourth-order algebraic equation for ω of the form $\omega^4 + b\omega^2 + c = 0$, with:

$$b = 4\pi G(\varrho_{B,0} + \varrho_{D,0}) - k^2(c_D^2 + c_B^2) \quad (A.39)$$

and

$$c = k^4 c_B^2 c_D^2 - 4\pi G(c_D^2 \varrho_{B,0} + c_B^2 \varrho_{D,0}) \quad (A.40)$$

The solutions will be of the type $\omega^2 = (-b \pm \sqrt{b^2 - 4c})/2$. We can see that if $c \geq 0$ we have $\sqrt{b^2 - 4c} < |b|$ so the sign of ω^2 will be the sign of $-b$. On the other hand if $c < 0$ then $\sqrt{b^2 - 4c} > |b|$ and there will be one positive and one negative solution for ω^2. We interpret negative solutions for ω^2 as instabilities as they represent exponential rather than oscillating solutions. Let us then study the sign of c. Dividing c by $k^2 c_D^2 c_B^2$ – a positive number – we find that the sign of c is the same as the sign of $k^2 - k_j^2$, where we have defined k_j^2 as:

$$k_j^2 = 4\pi G \varrho_{B,0} \left(1 + \frac{c_B^2 \varrho_{D,0}}{c_D^2 \varrho_{B,0}}\right) \quad (A.41)$$

Thus, we have recovered a criterion similar to the single-fluid Jeans instability. k_j^2 is very similar to the equivalent for Jeans and is modified by the ratio of densities and sound speeds. Let us see if other instabilities are possible when $k^2 > k_j^2$. In this case the sign of ω^2 is given by the sign of $-b$ but with simple algebra we can find that in the case $k^2 \geq k_j^2$ the expression for $-b$ is:

$$-b = 4\pi G \left(\varrho_{B,0} \frac{c_D^2}{c_B^2} + \varrho_{D,0} \frac{c_B^2}{c_D^2}\right) > 0 \quad (A.42)$$

Thus, the only instabilities are those corresponding to $k^2 < k_j^2$. Let us verify how different a result this is compared to the single-component Jeans instability for our unstable halos. Before any cooling $\varrho_{D,0} \simeq 5.78 \varrho_{B,0}$ from the ratio of the relative Ω values and the sound speeds are identical. However, gas can become Jeans unstable only after cooling and we have seen in Section 2.3.3 that this implies a temperature reduction by a factor ~ 6.97 and a similar increase in density. Thus, at the threshold of instability we have $\varrho_{D,0} \simeq 0.829 \varrho_{B,0}$ and – given that the sound speed squared scales with the temperature – $c_D^2 \simeq 6.97 c_B^2$. The additional term in the parenthesis of (A.41) is then $\simeq 0.119$. We can then conclude that using the two-fluid treatment instead of the traditional single-fluid Jeans instability affects the stability condition only at the $\sim 6\%$ level for the minimum unstable wavelength.

Bibliography

1 Abel, T., Anninos, P., Zhang, Y., Norman, M. L. *New Astronomy*, 2, 181, **1997**

2 Abel, T., Anninos, P., Norman, M. L., Zhang, Y. *Astrophysical Journal*, 508, 518, **1998**

3 Abel, T., Bryan, G. L., Norman, M. L. *Astrophysical Journal*, 540, 39, **2000**

4 Adelberger, K. L., Steidel, C. C., Shapley, A. E., Pettini, M. *Astrophysical Journal*, 584, 45, **2003**

5 Adelberger, K. L., Steidel, C. C., Kollmeier, J. A., Reddy, N. A. *Astrophysical Journal*, 637, 74, **2006**

6 Ahn, K., Shapire, P. R., Iliev, I. T., Mellema, G., Pen, U.-L. *Astrophysical Journal* submitted, see also arXiv:0807.2254

7 Aloisi, A., et al. *Astrophysical Journal*, 667, L151, **2007**

8 Anninos, P., Norman, M. L. *Astrophysical Journal*, 459, 12, **1996**

9 Anninos, P., Zhang, Y., Abel, T., Norman, M. L. *New Astronomy*, 2, 209, **1997**

10 Arendt, R. G., Fixsen, D. J., Moseley, S. H., *Astrophysical Journal*, 536, 500, **2000**

11 Baltz, E. A., Gnedin, N. Y., Silk, J. *Astrophysical Journal*, 493, L1, **1998**

12 Baraffe, I., Heger, A., Woosley, S. E. *Astrophysical Journal*, 550, 890, **2001**

13 Barnes, J. E., Hut, P. *Astrophysical Journal Supplement Series*, 70, 389, **1989**

14 Bechtold, J. *Astrophysical Journal Supplement Series*, 91, 1, **1994**

15 Bechtold, J., Dobrzycki, A., Wilden, B., Morita, M., Scott, J., Dobrzycka, D., Tran, K.-V., Aldcroft, T. L. *Astrophysical Journal Supplement Series*, 140, 143, **2002**

16 Becker, R. H., et al. *Astronomical Journal*, 122, 2850, **2001**

17 Becker, G. D., Rauch, M., Sargent, W. L. W. *Astrophysical Journal*, 662, 72, **2007**

18 Beckwith, S. V. W., et al. *Astronomical Journal*, 132, 1729, **2006**

19 Begelman, M. C., Volonteri, M., Rees, M. J. *Monthly Notices of the Royal Astronomical Society*, 370, 289, **2006**

20 *The Next Generation Space Telescope*, eds Bely, P.-Y., Burrows, C. J., Illingworth, G. D., STScI: Baltimore, **1989**

21 Benitez, N. *Astrophysical Journal*, 536, 571, **2000**

22 Bernardi, M. *Astronomical Journal*, 125, 32, **2003**

23 Bertin, E., Arnouts, S. *Astronomy & Astrophysics Supplement*, 117, 393, **1996**

24 Bertoldi, F., Carilli, C. L., Cox, P., Fan, X., Strauss, M. A., Beelen, A., Omont, A., Zylka, R. *Astronomy & Astrophysics*, 406, L55, **2003**

25 Bianchi, S., Cristiani, S., Kim, T.-S. *Astronomy & Astrophysics*, 376, 1, **2001**

From First Light to Reionization. Massimo Stiavelli
Copyright © 2009 WILEY-VCH Verlag GmbH & Co. KGaA, Weinheim
ISBN: 978-3-527-40705-7

26 Binney, J., Tremaine, S., *Galactic Dynamics*, 2nd edn, Princeton UP, Princeton, NJ, **2008**

27 Blanchard, A., Valls-Gabaud, D., Mamon, G. A. *Astronomy & Astrophysics*, 264, 365, **1992**

28 Bode, P., Ostriker, J. P., Turok, N. *Astrophysical Journal*, 556, 93, **2001**

29 Bond, J. R., Efstathiou, G. *Astrophysical Journal*, 285, L45, **1984**

30 Bonilha, J. R., Ferch, R., Salpeter, E. E., Slater, G., Noerdlinger, P. D. *Astrophysical Journal*, 233, 649, **1979**

31 Bonnor, W. B. *Monthly Notices of the Royal Astronomical Society*, 116, 351, **1956**

32 Bouwens, R. J., Illingworth, G. D., Blakeslee, J. P., Franx, M. *Astrophysical Journal*, 653, 53, **2006**

33 Bouwens, R. J., Illingworth, G. D. *Nature*, 443, 189, **2006**

34 Bouwens, R. J., Illingworth, G. D., Franx, M., Ford, H. *Astrophysical Journal*, 670, 928, **2007**

35 Bouwens, R. J., et al., *Astrophysical Journal*, submitted, **2008**, see also arXiv:0805.0593

36 Brammer, G. B., van Dokkum, P. G., Coppi, P. *Astrophysical Journal*, in press, **2008**, see also arXiv:0807.1533

37 Broadhurst, T. J., Taylor, A. N., Peacock, J. A. *Astrophysical Journal*, 438, 49, **1995**

38 Bromm, V., Coppi, P. S., Larson, R. B. *Astrophysical Journal*, 564, 23, **2002**

39 Bromm, V., Kudritzki, R. P., Loeb, A. *Astrophysical Journal*, 552, 464, **2001**

40 Bromm, V., Ferrara, A., Coppi, P.S., Larson, R. B. *Monthly Notices of the Royal Astronomical Society*, 328, 969, **2001**

41 Bromm, V., Clarke, C. J. *Astrophysical Journal*, 566, L1, **2002**

42 Brown, P. N., Byrne, G. D., Hindmarsh, A. C. *SIAM J. Sci. Stat. Comput.*, 10, 1038, **1989**

43 Bruzual, G., Charlot, S. *Monthly Notices of the Royal Astronomical Society*, 344, 1000, **2003**

44 Bunker, A. J., Stanway, E. R., Ellis, R. S., McMahon, R. G. *Monthly Notices of the Royal Astronomical Society*, 355, 374, **2004**

45 Cantalupo, S., Porciani, C., Lilly, S. J. *Astrophysical Journal*, 672, 48, **2008**

46 Carilli, C., in *From planets to dark ages: the modern radio Universe*, Proceedings of Science, ed Beswick, **2008**

47 Caroll, S. M., Press, W. H., Turner, E. L., *Ann. Rev. Astron. Astrophys.*, 30, 499, **1992**

48 Casertano, S., et al. *Astronomical Journal*, 120, 2747, **2000**

49 Castellani, V., Chieffi, A., Tornambe, A. *Astrophysical Journal*, 272, 249, **1983**

50 Chandrasekhar, S., *Principles of Stellar Dynamics*, Dover, New York, **1960**

51 Chandrasekhar, S., *An Introduction to the Study of Stellar Structure*, Dover, New York, **1967**

52 Charlot, S., Fall, S. M. *Astrophysical Journal*, 415, 580, **1993**

53 Chiu, W. A., Gnedin, N. Y., Ostriker, J. P. *Astrophysical Journal*, 563, 21, **2001**

54 Chokshi, A., Turner, E. L. *Monthly Notices of the Royal Astronomical Society*, 259, 421, **1992**

55 Christlieb, N., et al. *Nature*, 419, 904, **2002**

56 Christlieb, N., Gustafsson, B., Korn, A. J., Barklem, P. S., Beers, T. C., Bessel, M. S., Karlsson, T., Mizuno-Wiedner, M., *Astrophysical Journal*, 603, 708, **2004**

57 Ciardi, B., Madau, P. *Astrophysical Journal*, 596, 1, **2003**

58 Clark, P. C., Glover, S. C. O., Klessen, R. S. *Astrophysical Journal*, 672, 757, **2008**

59 Cooke, A. J., Espey, B., Carswell, R. F. *Monthly Notices of the Royal Astronomical Society*, 284, 552, **1997**

60 Cooray, A., Milosavljevic, M., *Astrophysical Journal*, 627, L89, **2005**

61 Couchman, H. M. P. *Monthly Notices of the Royal Astronomical Society*, 214, 137, **1985**

62 Couchman, H. M. P., Rees, M. J. *Monthly Notices of the Royal Astronomical Society*, 221, 53, **1986**

63 Couchman, H. M. P. *Astrophysical Journal*, 368, 23, **1991**

64 Croom, S. M., et al. *Astrophysical Journal*, 349, 1397, **2004**

65 Daub, C. T. *Astrophysical Journal*, 137, 184, **1963**

66 Davidsen, A. F., Kriss, G. A., Zheng, W. *Nature*, 380, 47, **1996**

67 Davidson, K., Kinman, T. D. *Astrophysical Journal Supplement Series*, 58, 321, **1985**

68 de Vaucouleurs, G., de Vaucouleurs, A., Corwin, H. G. Jr., Bura, R., Paturel, G., Fouque, P. *Third Reference Catalogue of Bright Galaxies*, Springer-Verlag: Berlin, **1991**

69 Dickinson, M., et al. *Astrophysical Journal*, 600, L99, **2004**

70 Dikstra, M., Haiman, Z., Rees, M. J., Weinberg, D. H. *Astrophysical Journal*, 601, 666, **2004**

71 Doroshkevich, A. G., Naselsky, I. P., Naselsky, P. D., Novikov, I. D. *Astrophysical Journal*, 586, 709, **2003**

72 Dove, J. B., Shull, J. M. *Astrophysical Journal*, 430, 222, **1994**

73 Dove, J. B., Shull, J. M., Ferrara, A. *Astrophysical Journal*, 531, 846, **2000**

74 *Exploration and the Search for Origins: A vision for ultraviolet-optical-infrared space astronomy*, Report of the 'HST and Beyond' Committee, ed A. Dressler, AURA: Washington, DC, **1996**

75 Ebert, R. *Zeitschrift für Astrophysik*, 37, 217 **1955** (in German)

76 Efstathiou, G., Eastwood, J. W. *Monthly Notices of the Royal Astronomical Society*, 194, 503, **1981**

77 Efstathiou, G. *Monthly Notices of the Royal Astronomical Society*, 256, P43 **1992**

78 Fall, S. M., Rees, M. J. *Astrophysical Journal*, 298, 18, **1985**

79 Fan, X. et al. *Astronomical Journal*, 128, 515, **2004**

80 Fan, X. et al. *Astronomical Journal*, 132, 117, **2006**

81 Fechner, C., et al. *Astronomy & Astrophysics*, 455, 91, **2006**

82 Ferland, G. J., Korista, K. T., Verner, D. A., Ferguson, J. W., Kingdon, J. B., Verner, E. M. *Publications of the Astronomical Society of the Pacific*, 110, 761, **1998**

83 Ferguson, H. C., Dickinson, M., Williams, R. *Ann. Rev. Astron. Astrophys.*, 38, 667, **2000**

84 Field, G. B. *Astrophysical Journal*, 129, 536, **1959**

85 Field, G. B., *Astrophysical Journal*, 129, 551, **1959**

86 Fixsen, D. J., Moseley, S. H., Arendt, R. G. *Astrophysical Journal Supplement Series*, 128, 651, **2000**

87 Fontanot, F., Cristiani, S., Monaco, P., Nonino, M., Vanzella, E., Brandt, W. N., Grazian, A., Mao, J. *Astronomy & Astrophysics*, 461, 39, **2007**

88 Frebel, A., Collet, R., Eriksson, K., Christlieb, N., Aoki, W. *Astrophysical Journal*, 684, 588, **2008**

89 Freudling, W., Corbin, M. R., Korista, K. T. 587, L67, **2003**

90 Fruchter, A. S., Hook, R. N. *Proceedings of the Astron. Soc. of the Pacific*, 114, 144, **2002**

91 Furlanetto, S. R., Sokasian, A., Hernquist, L. *Monthly Notices of the Royal Astronomical Society*, 347, 187, **2004**

92 Furlanetto, S. R., Zaldarriaga, M., Hernquist, L. *Astrophysical Journal*, 613, 16, **2004**

93 Fukugita, M., Hogan, C. J., Peebles, P. J. E. *Astrophysical Journal*, 503, 518, **1998**

94 Galli, D., Palla, F. *Astronomy & Astrophysics*, 335, 403, **1998**

95 Gao, L., White, S. D. M., Jenkins, A., Frenk, C. S., Springel, V. *Monthly Notices of the Royal Astronomical Society*, 363, 379, **2005**

96 Giallongo, E., Cristiani, S., D'Odorico, S., Fontana, A., Savaglio, S. *Astrophysical Journal*, 466, 46, **1996**

97 Giavalisco, M. *Ann. Rev. Astron. and Astrophys.*, 40, 579, **2002**

98 Giavalisco, M., et al. *Astrophysical Journal*, 600, L93, **2004**

99 Glover, S. C. O., Brand, P. W. J. L. *Monthly Notices of the Royal Astronomical Society*, 321, 385, **2001**

100 Gonçalves, T. S., Steidel, C. C., Pettini, M. *Astrophysical Journal*, 676, 816, **2008**

101 Gnedin, N. Y., Ostriker, J. P. *Astrophysical Journal*, 486, 581, **1997**

102 Greif, T. H., Johnson, J. L., Klessen, R. S., Bromm, V. *Monthly Notices of the Royal Astronomical Society*, submitted. See also arXiv:0803.2237

103 Gunn, J. E., Peterson, B. A. *Astrophysical Journal*, 142, 1633, **1965**

104 Haiman, Z., Thoul, A. A., Loeb, A. *Astrophysical Journal*, 464, 523, **1996**

105 Haiman, Z., Loeb, A. *Astrophysical Journal*, 483, 21, **1997**

106 Haiman, Z., Loeb, A. *Astrophysical Journal*, 484, 985, **1997**

107 Haiman, Z., Madau, P., Loeb, A. *Astrophysical Journal*, 514, 535, **1999**

108 Haiman, Z., Rees, M. J., Loeb, A. *Astrophysical Journal*, 484, 985, **1997**

109 Haiman, Z. *Astrophysical Journal*, 576, L1, **2002**

110 Halpern, J. P, Grindlay, J. E. *Astrophysical Journal*, 242, 1041, **1980**

111 Heap, S. R., Williger, G. M., Smette, A., Hubeny, I., Sahu, M., Jenkins, E. B., Tripp, T. M., Winkler, J. N. *Astrophysical Journal*, 534, 69, **2000**

112 Heckman, T. M., Robert, C., Leitherer, C., Garnett, D. R., van der Rydt, F. *Astrophysical Journal*, 503, 646, **1998**

113 Heger, A., Woosley, S. E. *Astrophysical Journal*, 567, 532, **2002**

114 Hernquist, L. *Astrophysical Journal Supplement Series*, 64, 715, **1987**

115 Hinshaw, G., et al. *Astrophysical Journal Supplement Series* submitted, **2008**, see also arXiv:0803.0732

116 Hu, E. M., Cowie, L. L., Capak, P., McMahon, R. G., Hayashino, T., Komiyama, Y. *Astronomical Journal*, 127, 563, **2004**

117 Iliev, I. T., et al. *Monthly Notices of the Royal Astronomical Society*, 371, 1057, **2006**

118 Iliev, I. T., Mellema, G., Pen, U.-L., Bond, J. R., Shapiro, P. R. *Monthly Notices of the Royal Astronomical Society*, 384, 863, **2008**

119 Jakobsen, P., Boksenberg, A., Deharveng, J. M., Greenfield, P., Jedrzejewski, R., Paresce, F. *Nature*, 370, 35, **1994**

120 Jappsen, A.-K., Glover, S. C. O., Klessen, R. S., Mac Low, M.-M. *Astrophysical Journal*, 660, 1332, **2007**

121 Jenkins, A. et al. *Astrophysical Journal*, 499, 20, **1998**

122 Jenkins, A. et al. *Monthly Notices of the Royal Astronomical Society*, 321, 372, **2001**

123 Johnson, J. L., Greif, T. H., Bromm, V. *Monthly Notices of the Royal Astronomical Society* in press, **2008** see also arXiv:0711.4622

124 Jones, B. J. T., Wyse, R. F. G. *Astronomy & Astrophysics*, 149, 144, **1985**

125 Kaplinghat, M., Chu, M., Haiman, Z., Holder, G. P., Knox, L, Skordis, C. *Astrophysical Journal*, 583, 24, **2003**

126 Kashlinksky, A., Arendt, R. G., Mather, J., Moseley, S. H. *Astrophysical Journal*, 666, L1, **2007**

127 Kippenhahn, R., Weigert, A. in *Stellar Structure and Evolution* (Heidelberg: Springer), 165

128 Koekemoer, A. M., *et al. HST Dither Handbook*, STScI: Baltimore, **2002**

129 Kogut, A., *et al. Astrophysical Journal Supplement Series*, 148, 161, **2003**

130 Komatsu, E., *et al. Astrophysical Journal Supplement Series* submitted, **2008**, see also arXiv:0803.0547

131 Kriss, G. A. *Science*, 293, 1112, **2001**

132 Kulkarni, V. P., Fall, S. M. *Astrophysical Journal*, 580, 732, **2002**

133 Laidler, V. G., Papovich, C., Grogin, N. A., Idzi, R., Dickinson, M., Ferguson, H. C., Hilbert, B., Clubb, K., Ravindranath, S. *Publ. Astronomical Society of the Pacific*, 119, 1325, **2007**

134 Larson, R. B. *Monthly Notices of the Royal Astronomical Society*, 145, 271 **1969**

135 Lee, K.-S., Giavalisco, M., Gnedin, O. Y., Somerville, R. S., Ferguson, H. C., Dickinson, M., Ouchi, M. *Astrophysical Journal*, 642, 63, **2006**

136 Leitherer, C., Ferguson, H. C., Heckman, T. M., Lowenthal, J. D. *Astrophysical Journal*, 454, L19, **1995**

137 Lepp, S., Shull, J. M. *Astrophysical Journal*, 280, 465, **1984**

138 Lidz, A., Zahn, O., McQuinn, M., Zaldarriaga, M., Hernquist, L. *Astrophysical Journal*, 680, 962, **2008**

139 Lidz, A., Zahn, O., Furlanetto, S. R., McQuinn, M., Hernquist, L., Zaldarriaga, M. *Astrophysical Journal* submitted, **2008**, see also arXiv:0806.1055

140 Loeb, A., Rybicki, G. B. *Astrophysical Journal*, 524, 527, **1999**

141 Loeb, A., Barkana, R. *Annual Reviews of Astronomy and Astrophysics*, 39, 19, **2001**

142 Loeb, A., Zaldarriaga, M. *Phys. Rev. Lett*, 92, 1301, **2004**

143 Lotz, J. M., Telford, R., Ferguson, H. C., Miller, B. W., Stiavelli, M., Mack, J. *Astrophysical Journal*, 552, 572, **2001**

144 Lu, L., Wolfe, A. M., Turnsekh, D. A. *Astrophysical Journal*, 367, 19, **1991**

145 Lucy, L. B. *Astronomical Journal*, 79, 645, **1974**

146 Lucy, L. B. *Astronomy & Astrophysics*, 261, 706, **1992**

147 Lucy, L. B. *Astronomical Journal*, 104, 1260, **1992**

148 Machacek, M. E., Bryan, G. L., Abel, T. *Astrophysical Journal*, 548, 188, **2001**

149 Machida, M. N., Matsumoto, T., Inutsuka, S. *Astrophysical Journal* submitted, **2008**. See also arXiv:0803.1224

150 MacKey, J., Bromm, V., Hernquist, L. *Astrophysical Journal*, 586, 1, **2003**

151 Madau, P., Meiksin, A. *Astrophysical Journal*, 433, L53, **1994**

152 Madau, P. 1995, 441, 18, **1995**

153 Madau, P., Ferguson, H. C., Dickinson, M. E., Giavalisco, M., Steidel, C. C., Fruchter, A. *Monthly Notices of the Royal Astronomical Society*, 283, 1388, **1996**

154 Madau, P., Shull, J. M. *Astrophysical Journal*, 457, 551, **1996**

155 Madau, P., Meiksin, A., Rees, M. J. *Astrophysical Journal*, 475, 429, **1997**

156 Madau, P., Haardt, F., Rees, M. J. *Astrophysical Journal*, 514, 648, **1999**

157 Madau, P., Rees, M. J. *Astrophysical Journal*, 542, L69, **2000**

158 Madau, P., Ferrara, A., Rees, M. J. *Astrophysical Journal*, 555, 82, **2001**

159 Madau, P., Rees, M. J. *Astrophysical Journal*, 551, L27, **2001**

160 Malhotra, S., Rhoads, J. E. *Astrophysical Journal*, 617, L5, **2004**

161 Malhotra, S., et al. *Astrophysical Journal*, 626, 666, **2005**

162 Malhotram S., Rhoads, J. E. *Astrophysical Journal*, 647, L95, **2006**

163 Marigo, P., Girardi, L., Chiosi, C., Wood, P. R. *Astronomy & Astrophysics*, 371, 152, **2001**

164 Meiksin, A., *Monthly Notices of the Royal Astronomical Society*, 365, 807, **2006**

165 Merritt, D., Ferrarese, L. *Monthly Notices of the Royal Astronomical Society*, 320, L30, **2001**

166 McKee, C., Taylor, J., *Astronomy and Astrophysics in the New Millenium*, National Academ of Sciences, Washington, **2001**

167 Mezger, P. G. *Astronomy & Astrophysics*, 70, 565, **1978**

168 Miralda-Escude, J., Rees, M. J. *Monthly Notices of the Royal Astronomical Society*, 266, 343, **1994**

169 Miralda-Escude, J. *Astrophysical Journal*, 501, 15, **1998**

170 Miralda-Escude, J., Rees, M. J. *Astrophysical Journal*, 542, L69, **1998**

171 Miralda-Escude, J., Haehnelt, M., Rees, M. J. *Astrophysical Journal*, 530, 1, **2000**

172 Mobasher, B., et al. *Astrophysical Journal*, 635, 832, **2005**

173 Mobasher, B., et al. *Astrophysical Journal Supplement Series*, 172, 117, **2007**

174 Møller, P., Kjaergaard, P. *Astronomy & Astrophysics*, 258, 234, **1992**

175 Moore, B. *Astrophysical Journal*, 461, L13, **1996**

176 Moore, B., Diemand, J., Madau, P., Zemp, M., Stadel, J. *Monthly Notices of the Royal Astronomical Society*, 368, 563, **2006**

177 Mori, M., Ferrara, A., Madau, P. *Astrophysical Journal*, 571, 40, **2002**

178 Mortonson, M. J., Hu, W. *Astrophysical Journal* submitted, **2008**, see also arXiv:0804.2631

179 Mukherjee, P., Liddle, A. R. *Monthly Notices of the Royal Astronomical Society*, 389, 231, **2008**

180 Muñoz, J. A., Loeb, A., *Monthly Notices of the Royal Astronomical Society*, 385, 2175, **2008**

181 Mutchler, M., Sirianni, M., van Orsow, D., Riess, A. *Bias and dark calibration of ACS data*, ISR ACS 2004-07, STScI:Baltimore, **2004**

182 Navarro, J. F., Frenk, C. S., White, S. D. M. *Astrophysical Journal*, 490, 493, **1997**

183 Oh, S. P. *Monthly Notices of the Royal Astronomical Society*, 336, 1021, **2002**

184 Oke, J. B., Gunn, J. *Astrophysical Journal*, 266, 713, **1983**

185 Oesch, P. A., et al. *Astrophysical Journal*, 671, 1212, **2007**

186 Oesch, P. A., et al. *Astrophysical Journal* in press, see also arXiv:0804.4874

187 Oh, S. P., Nollett, K. M., Madau, P., Wasserburg, G. J. *Astrophysical Journal*, 562, L1, **2001**

188 Omukai, K., Palla, F. *Astrophysical Journal*, 561, L55, **2001**

189 Omukai, K., Palla, F. *Astrophysical Journal*, 589, 677, **2003**

190 O'Shea, B. W., Bryan, G., Bordner, J., Norman, M. L., Abel, T., Harkness, R., Kritsuk, A, in *Adaptive Mesh Refinement – Theory and Application*, eds Plewa, Linde and Weirs, Springer Lecture notes in Computational Science and Engineering, **2004**, see also arXiv:astro-ph/0403044

191 O'Shea, B. W., Abel, T., Whalen, D., Norman, M. L., *Astrophysical Journal*, 628, L5, **2005**

192 O'Shea, B. W., Norman, M. L., *Astrophysical Journal*, 648, 31, **2006**

193 O'Shea, B. W., Norman, M. L., *Astrophysical Journal*, 654, 66, **2007**

194 O'Shea, B. W., Norman, M. L., *Astrophysical Journal*, 673, 14, **2008**

195 O'Shea, B. W., McKee, C. F., Heger, A., Abel, T. in *Proceedings of First Stars III*, eds B. W. O'Shea, A. Heger, and T. Abel, **2008**, see also arXiv:0801.2124

196 Osterbrock, D. E. *Astrophysics of Gaseous Nebulae and Active Galactic Nuclei* (Mill Valley, CA: University Science Books), **1989**

197 Osterbrock, D. E., Ferland, G. J. *Astrophysics of Gaseous Nebulae and Active Galactic Nuclei* (Sauslito, CA: University Science Books), **2005**

198 Ostriker, J. P., Gnedin, N. Y. *Astrophysical Journal*, 472, L63, **1996**

199 Ota, K., Kashikawa, N., Malkan, M. A., Iye, M., Nakajima, T., Nagao, T., Shimasaku, K., Gandhi, P. *Astrophysical Journal*, 677, 12, **2008**

200 Pagel, B. E. J. 2000, *Physical Review*, 333, 433, **2000**

201 Panagia, N. *Astronomical Journal*, 78, 929, **1973**

202 Panagia, N., Ranieri, M. *Mem. Soc. R. Sci. Liege*, 6(V), 275, **1973**

203 Panagia, N., Stiavelli, M., Ferguson, H., Stockman, H. S. *Revisa Mexicana de Astronomia y Astrofisica*, 17, 230, **2003**

204 Pascarelle, S. M., Lanzetta, K. M., Chen, H.-W., Webb, J. K. *Astrophysical Journal*, 560, 101, **2001**

205 Peacock, J. A. *Cosmological Physics*, Cambridge U. Press, **1999**

206 Peebles, P. J. E. *Principles of Physical Cosmology*, Princeton U. Press, **1993**

207 Peebles, P. J. E. *Astrophysical Journal*, 153, 1, **1968**

208 Peebles, P. J. E., Dicke, R. H. *Astrophysical Journal*, 154, 891, **1968**

209 Pei, Y. C., Fall, S. M., Hauser, M. G. *Astrophysical Journal*, 572, 604, **1999**

210 Pentericci, L., Rix, H.-W., Prada, F., Fan, X., Strauss, M. A., Schneider, D. P., Grebel, E. K., Harbeck, D., Brinkmann, J., Narayan, V. K. *Astronomy & Astrophysics*, 410, 75, **2003**

211 Popa, L. A., et al. *New Astronomy Reviews*, 51, 298, **2007**

212 Porciani, C., Giavalisco, M. *Astrophysical Journal*, 565, 24, **2002**

213 Postman, M. *Proc. of SPIE*, 7010, 701021, **2008**

214 Press, W. H., Schechter, P. *Astrophysical Journal*, 187, 425, **1974**

215 Press, W. H., Teukolsky, S. A., Vetterling, W. T., Flannery, B. P. *Numerical Recipes in Fortran: The Art of Scientific Computing*, Cambridge Univ. Press, **1993**

216 Press, W. H., Teukolsky, S. A., Vetterling, W. T., Flannery, B. P. *Numerical Recipes in C: The Art of Scientific Computing*, Cambridge Univ. Press, **1993**

217 Prochaska, J. X., Wolfe, A. M., Tyytler, D., Burles, S., Cooke, J., Gawiser, E., Kirkman, D., O'Meara, J. M., Storrie-Lombardi, L. *Astrophysical Journal Supplement Series*, 137, 21, **2001**

218 Purcell, E. M., Field, G. B. *Astrophysical Journal*, 124, 542, **1956**

219 Quinn, T., Katz, N., Efstathiou, G. *Monthly Notices of the Royal Astronomical Society*, 278, L49, **1996**

220 Raiter, A., Fosbury, R., Stiavelli, M. in *Far away light in the young Universe at redshift beyond 3*, http://www.iap.fr/col2008/Proceedings/POSTERS/Raiter_poster_AP2008.pdf

221 Reed, D. S., Bower, R., Frenk, C. S., Jenkins, A., Theuns, T. *Monthly Notices of the Royal Astronomical Society*, 374, 2, **2007**

222 Rees, M. in *The Next Generation Space Telescope: Science Drivers and Technological Challenges*, Kaldeich ed, ESA SP 429, 5, **1998**

223 Reimers, D., Fechner, C., Hagen, H.-J., Jakobsen, P., Tytler, D., Kirkman, D. *Astronomy & Astrophysics*, 442, 63, **2005**

224 Rhoads, J. E., Malhotra, S. *Astrophysical Journal*, 563, L5, **2001**

225 Rhoads, J. E., Xu, C., Dawson, S., Dey, A., Malhotra, S., Wang, J.-X., Jannuzi, B. T., Spinrad, H., Stern, D. *Astrophysical Journal*, 2004, 611, 59, **2004**

226 Richards, G. T., et al., *Monthly Notices of the Royal Astronomical Society*, 360, 839, **2005**

227 Ricotti, M., Shull, J. M. *Astrophysical Journal*, 542, 548, **2000**

228 Roser, S., Bastian, U. *Bull. Inf. Cent. Donnees Astron. Strasbourg*, 42, 11, **1993**

229 Saigo, K., Matsumoto, T., Umemura, M. *Astrophysical Journal*, 615, L65, **2004**

230 Santos, M. R. *Monthly Notices of the Royal Astronomical Society*, 349, 1137, **2004**

231 Schaerer, D. *Astronomy & Astrophysics*, 382, 28, **2002**

232 Schaerer, D. *Astronomy & Astrophysics*, 397, 527, **2003**

233 Schechter, P. *Astrophysical Journal*, 203, 297, **1976**

234 Scott, J., Bechtold, J., Dobrzycki, A., Kulkarni, V. P. *Astrophysical Journal Supplement Series*, 130, 67, **2000**

235 Seager, S., Sasselov, D. D., Scott, D. *Astrophysical Journal Supplement Series*, 128, 407, **2000**

236 Seager, S., Sasselov, D. D., Scott, D. *Astrophysical Journal*, 523, L1, **1999**

237 Shapley, A. E., Steidel, C. C., Pettini, M., Adelberger, K. L., Erb, D. K. *Astrophysical Journal*, 651, 688, **2006**

238 Shaver, P.A., Windhorst, R.A., Madau, P., de Bruyn. A. G. *Astronomy & Astrophysics*, 345, 380, **1999**

239 Sheth, R. K., Tormen, G. *Monthly Notices of the Royal Astronomical Society*, 329, 61, **2002**

240 Shull, J. M. *Astrophysical Journal*, 234, 761, **1979**

241 Shull, J. M., Beckwith, S. *Ann. Rev. Astron. and Astrophys.*, 20, 163, **1982**

242 Silk, J., *Astrophysical Journal*, 151, L19, **1968**

243 Silk, J., Rees, M. J. *Astronomy & Astrophysics*, 331, L1, **1998**

244 Smette, A., Heap, S. R., Willigerm G. M., Tripp, T. M., Jenkins, E. B., Songaila, A. *Astrophysical Journal*, 564, 542, **2002**

245 Smith, B. D., Sigurdsson, S. *Astrophysical Journal*, 661, L5, **2007**

246 Soifer, B. T., Neugebauer, G., Helou, G., Lonsdale, C. J., Hacking, P., Rice, W., Houck, J. R., Low, F. J., Rowan-Robinson, M. *Astrophysical Journal*, 283, L1, **1984**

247 Songaila, A. *Astrophysical Journal*, 561, L153, **2001**

248 Somerville, R. S., Lee, K., Ferguson, H. C., Gardner, J. P., Moustakas, L. A., Giavalisco, M. *Astrophysical Journal*, 600, L171, **2004**

249 Songaila, A. *Astronomical Journal*, 115, 2184, **1998**

250 Songaila, A. *Astronomical Journal*, 130, 1996, **2005**

251 Spergel, D. N. et al. *Astrophysical Journal Supplement Series*, 148, 175, **2003**

252 Spergel, D. N., et al. *Astrophysical Journal Supplement Series*, 17, 377, **2007**

253 Springel, V. *Monthly Notices of the Royal Astronomical Society*, 364, 1105, **2005**

254 Stacy, A., Bromm, V. *Monthly Notices of the Royal Astronomical Society*, 382, 229, **2007**

255 Stecher, T. P., Williams, D. A., *Astrophysical Journal*, 149, L29, **1967**

256 Steidel, C. C., Adelberger, K. L., Giavalisco, M., Dickinson, M., Pettini, M. *Astrophysical Journal*, 519, 1, **1999**

257 Steidel, C. C., Adelberger, K. L., Shapley, A. E., Pettini, M., Dickinson, M., Giavalisco, M. *Astrophysical Journal*, 532, 170, **2000**

258 Steidel, C. C., Pettini, M., Adelberger, K. L. *Astrophysical Journal*, 546, 665, **2001**

259 Steidel, C. C., Adelberger, K. L., Shapley, A. E., Pettini, M., Dickinson, M., Giavalisco, M. *Astrophysical Journal*, 592, 728, **2003**

260 Stern, D., Yost, S. A., Eckart, M. E., Harrison, F. A., Helfand, D. J., Djrgovski, S. g., Malhotra, S., Rhoads, J. E. *Astrophysical Journal*, 619, 12, **2005**

261 Stiavelli, M., White, R. L. *Data Compression for ACS*, ISR ACS 97-02, **1997**

262 Stiavelli, M., et al. *JWST Primer*, STScI: Baltimore, **2008**

263 Stiavelli, M. *Astrophysical Journal*, 495, L91, **1998**

264 Stiavelli, M., Fall, S. M., Panagia, N. *Astrophysical Journal*, 600, 508, **2004**

265 Stiavelli, M., Fall, S. M., Panagia, N. *Astrophysical Journal*, 610, L1, **2004**

266 Stiavelli, M., Djorgovski, S. G., Pavlovsky, C., Scarlata, C., Stern, D., Mahabal, A., Thompson, D., Dickinson, M., Panagia, N., Meylan, G. *Astrophysical Journal*, 622, L1, **2005**

267 Stone, J. M., Norman, M. L. *Astrophysical Journal Supplement Series*, 80, 753, **1992**

268 *The Next Generation Space Telescope: Visiting a time when galaxies were young*, Ed. H. S. Stockman, AURA: Washington, DC, **1997**

269 Sutherland, R. S., Dopita, M. A. *Astrophysical Journal Supplement Series*, 88, 253, **1993**

270 Taniguchi, Y., *Publ. Astron. Soc. Japan*, 57, 165, **2005**

271 Tegmark, M., Silk, J., Rees, M. J., Blanchard, A., Abel, T., Palla, F. *Astrophysical Journal*, 474, 1, **1997**

272 Theuns, T., Schaye, J., Zaroubi, S., Kim, T.-S., Tzanavaris, P., Carswell, B. *Astrophysical Journal*, 567, L103, **2002**.

273 Thompson, R. I., Eisenstein, D., Fan, X., Rieke, M., Kennicutt, R. C. *Astrophysical Journal*, 666, 658, **2007**

274 Tittley, E. R., Meiksin, A. *Monthly Notices of the Royal Astronomical Society*, 380, 1369, **2007**

275 Tornatore, L., Ferrara, A., Schneider, R. *Monthly Notices of the Royal Astronomical Society*, 382, 945, **2007**

276 Tran, K.-V. H., Lilly, S. J., Crampton, D., Brodwin, M. *Astrophysical Journal*, 612, L89, **2004**

277 Trenti, M., Hut, P. *Gravitational N-body Simulations*, http://www.scholarpedia.org/article/N-nody_simulations, see also arXiv:0806.3050

278 Trenti, M., Stiavelli, M. *Astrophysical Journal*, 667, 38, **2007**

279 Trenti, M., Stiavelli, M. *Astrophysical Journal*, 676, 767, **2008**

280 Trenti, M., Santos, M. R., Stiavelli, M. *Astrophysical Journal*, in press **2008**

281 Tumlinson, J., Shull, J. M. *Astrophysical Journal*, 528, L65, **2000**

282 Tumlinson, J., Giroux, M. L, Shull, J. M. *Astrophysical Journal*, 550, L1, **2001**

283 Tumlinson, J., Venkatesan, A., Shull, J. M. *Astrophysical Journal*, 612, 602, **2004**

284 Vale, A., Ostriker, J. P. *Monthly Notices of the Royal Astronomical Society*, 353, 189, **2004**

285 Vanzella, E. *et al.* *Astronomy & Astrophysics*, 478, 83, **2008**

286 Veron-Cetty, P.-P., Veron, P. *Astronomy & Astrophysics*, 374, 92, **2001**

287 Villanova, S., Piotto, G., King, I. R., and 9 other authors *Astrophysical Journal*, 663, 296, **2007**

288 Vogel, S. N., Weymann, R., Rauch, M., Hamilton, T. *Astrophysical Journal*, 441, 162, **1995**

289 Volonteri, M., Rees, M. J. *Astrophysical Journal*, 650, 669, **2006**

290 Weinmann, S. M., Lilly, S. J. *Astrophysical Journal*, 624, 526, **2005**

291 Whalen, D., Abel, T., Norman, M. L. *Astrophysical Journal*, 610, 14, **2004**

292 White, R. L., Becker, R. H., Helfand, D. J., Gregg, M. D. *Astrophysical Journal*, 475, 479, **1997**

293 Whitmore, B. C., Chandar, R., Fall, S. M. *Astronomical Journal*, 133, 1067, **2007**

294 Wiklind, T., Dickinson, M., Ferguson, H. C., Giavalisco, M., Mobasher, B., Grogin, N. A., Panagia, N. *Astrophysical Journal*, 676, 781, **2008**

295 Wiliams, R. E., *et al.* *Astronomical Journal*, 112, 1335, **1996**

296 Williams, R. E., *et al.* *Astronomical Journal*, 120, 2735, **2000**

297 Wise, J. H., Abel, T. *Astrophysical Journal*, 671, 1559, **2007**

298 Wise, J. H., Abel, T. *Astrophysical Journal*, 685, 40, **2008**

299 Wolf, C., *et al.*, *Astronomy & Astrophysics*, 408, 499, **2003**

300 Woosley, S. E., Weaver, T. A. *Astrophysical Journal Supplement Series*, 101, 181, **1995**

301 Wouthuysen, S. A. *Astrophysical Journal*, 57, 31, **1952**

302 Wu, Y. *et al.* *Astrophysical Journal*, 662, 952, **2007**

303 Wyithe, J. S. B., Loeb, A., Schmidt, B. P. *Monthly Notices of the Royal Astronomical Society*, 380, 1087, **2007**

304 Wyithe, J. S. B., Loeb, A., Geil, P. M. *Monthly Notices of the Royal Astronomical Society*, 383, 1195, **2008**

305 Yan, H., Windhorst, R. A., Odewahn, S. C., Cohen, S. H., R ottgering, H. J. A., Keel, W. C. *Astrophysical Journal*, 580, 725, **2002**

306 Yan, H., Windhorst, R. A. *Astrophysical Journal*, 612, L93, **2004**

307 Yoshida, M., *et al.* *Astrophysical Journal*, 653, 988, **2006**

308 Youshida, N., Omukai, K, Hernquist, L. *Science*, 321, 669, **2008**

309 Zaldarriagam M., Furlanetto, S. R., Hernquist, L. *Astrophysical Journal*, 608, 622, **2004**

310 Zepf, S. E. *New Astronomy Reviews*, 49, 413, **2005**

311 Zheng, W., *et al.* *Astronomical Journal*, 127, 656, **2004**

312 Zheng, W. *et al.* *Astrophysical Journal* in preparation, **2008**

Index

symbols
ΛCDM   21, 100
[OIII]λ5007/Hβ ratio   69
21-cm   4ff, 121ff, 124, 126, 191
21-cm radiation   103

a
Abel   25
absorption cross section   93
accretion disk   40
accretion flow   28, 40
advanced camera for surveys   132
advanced technology large-aperture space telescope (ATLAST)   188
AGN   7, 9, 76, 80, 81
  -mini   7ff
ALMA   153, 181
amplification   167
angular power spectrum   123
antenna temperature   122ff, 123
AP3M codes   70
Atacama large millimiter array (ALMA)   194
attenuation length   99
Auger electrons   100
automated photometry   162

b
background
   -UV   1
background-limited   188
bad pixels   154
Balmer line   118
Balmer-jump technique   129, 136
Baltz   127
baryon acoustic oscillations   125
baryonic mass   27
baryons   7
Bechtold   104
Bergeron   156

Big Bang   2, 8
black hole   1, 7ff, 8, 9ff, 34, 64ff, 75ff
black-hole mergers   65
blackbody   42, 97
blackbody spectrum   33ff
block-averaging   162
Bonnor Ebert   27, 30
Bremsstrahlung   51
Bromm   33, 58, 60
Bruzual and Charlot   130
Bruzual–Charlot models   63

c
case A   51
case B   10, 36, 53
CDM   2, 37
  -power spectrum   7
CDM halo   64
central pressure   31
Chandra deep field south   153
Chandrasekhar   200
charge neutrality   31
chemical potentials   198
Cloudy   52, 67, 87
clumpiness   143
clumpiness factor   88
clumpiness parameter   98
clumpy IGM   88
clusters of Population III stars   61
CMB   1, 4, 121
CMB fluctuations   126
CMB temperature   122
CNO cycle   32
Colberg   167
cold dark matter   75
collisional effects   36
collisional excitation cooling   52
collisional excitation rate   120
collisional excitations   121
collisional ionization cooling   52

From First Light to Reionization. Massimo Stiavelli
Copyright © 2009 WILEY-VCH Verlag GmbH & Co. KGaA, Weinheim
ISBN: 978-3-527-40705-7

collisionless   68
collisionless dynamics   68
color–color diagram   130
column density   41, 43, 49
completeness   165
Compton optical depth   126
continuity equations   71, 201
cooling   8, 22, 23, 25
  -atomic-hydrogen   48
  -molecular-hydrogen   9, 45, 61
cooling function   16, 16ff, 22, 58, 61, 97
  -approximation   17
cooling luminosity   149
cooling of a dark halo   119
cooling time   59
cooling timescale   22, 97
corrective factor   11
correlated noise   159, 162
correlation $R(\theta)$ 147
cosmic microwave background   10ff, 97ff
cosmic microwave radiation   103
cosmic origins spectrograph   186
cosmic rays   40
cosmic variance   166, 168
cross section   94
cyrrus   152

d

damped Lyman α systems   97
damping wing   94, 106ff, 119
Dark Ages   1, 3ff
dark energy   125
dark gaps   110
dark matter   37
dark matter halos   8ff, 21
  -virialized   21ff
dark-gap size   110
deconvolution procedure   159
density   9
density contrast   18ff
density fluctuation   1, 8
diffuse neutral hydrogen   108
diffuse neutral medium   116
direct collapse   67
dispersion relation   202
dithering   154
Doppler core   94
Doyon   185
drizzle algorithm   158
drizzling   157
Drop size kernel function   158

dust   93
dust absorption   87
DVODE   11
dynamo amplification   31

e

Eddington luminosity   30, 31ff, 65, 149
effective absorption cross section   111
effective temperature   31ff, 42
Einstein–de Sitter   18ff
electron temperature   79
electrons   8ff, 39, 50
Enzo   72
equilibrium temperature   52, 53
equipartition   50
escape fraction   86, 87, 93, 97, 112, 115, 116, 118, 179
escape of ionizing photons   85
escape velocity   54, 56
ESO/VLT   137
Euler's equation   72, 200, 201
Eulerian Codes   72
extended halos   157
extended wings   147
extension threshold   162

f

Fall   84
fast Fourier transforms   70
feedback   28, 57
  -chemical   39
  -mechanical   39
  -radiative   39ff
field choice   151
filtering   156
finding objects   162
first active galactic nuclei   39, 64
First catalog   153
first galaxies   39, 39ff, 63, 129
first light   1, 7
  -stars   2
first star   7ff, 8, 18, 31, 40, 49, 148
first star clusters   39, 59
flat fielding   156
flat-fielding errors   155
fluctuations   147
fluctuations in the 21-cm line   122
formation timescale   45
fragmentation   40, 60
Fruchter and Hook   157

g

Gadget 72
galactic dust extinction 151
galaxies 1ff
   -first 4
Galli and Palla 14, 23
gas dynamics 71
gas temperature 121, 122
Gaussian 111
George Rieke 185
Giavalisco 130
Glover and Brand 42
Gnedin and Ostriker 88, 89
GOODS 92, 133, 141, 142
gravitational cross section 70
gravitational lensing 166ff
gravitational telescope 166
great observatories origins deep survey 151
Greif 61
Gunn–Peterson optical 77
Gunn–Peterson optical depth 174
Gunn–Peterson optical thickness 106
Gunn–Peterson trough 4ff, 76, 103ff, 106ff, 107, 114, 117, 173ff, 175

h

hardness 80
He 1700 + 6416 173
heating 50
Heger and Woosley 35
helium 16
   -neutral 8
herringbone artefact 156, 159
HI Lyman α forest 177
hierarchical clustering 23
HII region 35ff, 67, 86, 93, 97, 109
HII region size 110
homogeneous IGM 81
HS 1700+6416 174
HST 127
Hubble deep field *see also* HDF 130, 151
Hubble deep field south 151
Hubble space telescope *see also* HST 91, 130, 151, 181
Hubble time 18, 98
Hubble ultra deep field 63, 151, 152
HUDF 133, 141
Hutchings 185
hydrogen 9, 16
   -molecular 8ff, 13ff, 15, 23, 39
   -neutral 1, 4ff, 8, 39
   -positively ionized 8

i

i-dropout criterion 132
IGM 79ff
IGM absorption 87
IGM attenuation 131
IGM temperature 176
IMF 84
initial mass function *see also* IMF 8, 82, 112
inter stellar medium *see also* ISM 9
intergalactic medium *see also* IGM 9, 96
inverse Compton cooling 52, 97ff, 98
ionization 113
     -efficiency 81
     -rate 113
ionization fraction 10, 10ff, 89, 117
ionization front 55ff
ionization heating 49, 56
ionized bubble 116, 117
isobaric 25
isobaric approximation 27, 59

j

J033238.7-274839.8 137ff
Jakobsen 184
James Webb space telescope 3, *see also* JWST 91, 181
Jeans 21
Jeans collapse 27, 40, 59
Jeans instability 25, 59, 197
Jeans mass 26, 41, 59
Jeans wavelength 26
Jenkins 20
Jones and Wyse 11
JWST 127

k

Keck 137

l

Lagrangian codes 72
Landau damping 201
Lane–Emden 200
Lane–Emden equation 197
large field of view imaging *see also* LSST 194
last scattering surface 13, 126
Legendre transform 198

Index

Lepp and Shull 13, 15
lifetime 32
linear perturbations 2
local ionized bubble 114
Lockman hole 153
LOFAR 3, 181
low-frequency array *see also* LOFAR 192
luminosity function 91, 92, 118, 140, 141, 146, 168
luminosity–surface density relations 90
LW bands 41
Lyα equivalent width 68
Lyman
    -absorption 4
    -emitters 4
Lyman α 95
Lyman α absorbers 106
Lyman α emitters 145
Lyman α escape fraction 114, 115
Lyman α luminosity 113
Lyman α 4, 11, 36ff, 67, 76, 92ff, 103ff, 118, 174
Lyman α absorption systems 175
Lyman α blobs 120
Lyman α cooling 52
Lyman α cooling halo 120
Lyman α emitters 113, 125, 144
Lyman α escape model 115
Lyman α excess 129, 135
Lyman α fluorescence 119, 120
Lyman α forest absorbers 35
Lyman α halos 119
Lyman α line 127
Lyman α line width 111
Lyman α sources 111, 115, 117
Lyman β 108
Lyman γ 108
Lyman limit 49
Lyman series lines 108
Lyman–Werner background 39, 44ff, 45–48, 61, 65
Lyman–Werner continuum 28
Lyman–Werner feedback 148, 149
Lyman–Werner resonances 40
Lyman-break galaxies 134, 141ff, 142
Lyman-break technique 4, 129, 129ff

m

Machacek 46
Madau 99, 131
Madau and Meiksin 177
magnitude per isophotal 160
Malhotra and Rhoads 117
mass function 18
mean excess energy 50, 53
mean weighted excess energy 50, 53
mesh refinement 73
metal abundance ratios 63
metal lines 109
metal-line ratios 176
metal-poor stars 61
metallicity 2, 7, 36
metallicity at reionization 75, 83
minimum halo mass 25, 27, 43ff, 46, 53–55
minimum metallicity 61, 84
minimum molecular hydrogen fraction 46
minimum surface brightness 81, 84, 90
Miralda-Escudé and Rees 93
MIRI 185
molecular hydrogen
    -dissociation 43
    -primordial 43
molecular hydrogen fraction 23ff, 59
Moon radiotelescope 194
multiplicity condition 60
Murchison wide-field array *see also* MWA 192
MWA 181

n

N-body simulation 167
Navarro, Frenk and White 22
near-IR background 127
neutral hydrogen clouds 105
neutral medium 105, 106
neutral-hydrogen fraction 110
neutral-hydrogen searches 121
NICMOS 133, 158
NIRCam 184
NIRSpec 184
number density
    -dark halos 23
    -electrons 10
    -hydrogen 45
    -neutral (atomic) hydrogen 10
    -neutral-hydrogen 107
    -Population III 43, 44, 47, 65
    -protons 10
numerical simulations 3
Nyquist sample 187

o

O'Shea 40
odd–even pattern 63
OI 109
Omega Cen 62
opening angle 71
optical depth 42, 177
optical thickness 95
origin of globular clusters 62
Osterbrock 50
Osterbrock and Ferland 36, 79

p

P3M codes 70
pair-instability supernova 56, 64, 148
Panagia 36ff, 67
particle-mesh codes 70
partition function 199
Peacock 8
Peebles 8, 10ff
Peebles and Dicke 62
percolation 76
perfect gas equation of state 72
Perlmutter 195
perturbations
 -collapse 21
photoionization
 -helium 8
 -hydrogen 8
photoionization cross section 99
photometric error 162
photometric redshifts 139
2-photon 11, 36
Planck 195
Planck's constant 78
PM code 70
point spread function 155
Poisson's equation 69ff, 200, 201
polytropic equation of state 72
polytropic stars 200
Population 9
Population II.5 9, 40, 58
Population III 4ff, 7ff, 9ff, 18, 27ff, 31ff, 36, 39, 41, 55, 56, 81, 148ff, 186ff
Population III.1 39, 61
Population III.2 39
power spectrum 20, 124–126
 -fluctuations 124
PPM 153
Press–Schechter 20
primordial metallicity 48
primordial nucleosynthesis 9
protons 8ff, 39
protostellar core 28ff
proximity effects 107

q

QSO 8, 66ff, 103ff, 108, 129, 146, 173
quasars 129

r

radiation
 -ionizing 1
radiation constant 12
radiative equilibrium 51
radiative feedback 31, 49, 61, 65
radiative transfer 44, 47, 73
RC3 153
recombination 8, 10, 50, 121
 -helium 9
 -hydrogen 9ff
recombination coefficient 51
recombination cooling 51
recombination rate 79
recombination timescale 98
redshift 6 dropout galaxies 132
redshift distribution 132, 134, 135, 138–140
redshift histogram 144
reheating 97
reionization 9, 75, 79ff, 95, 96, 117
 -helium 75, 173
 -hydrogen 1ff, 75, 103ff
reionization history 103
remnants 34
residual ionization 12ff, 25
residual ionized fraction 46
Rice algorithm 156
Rieke, M. 184
Runge–Kutta 200

s

Saha's equation 10, 197, 198
Salpeter time 65
Santos 93
scattering halo 119
Schaerer 112
SDSS 108
Seager 12ff
self-calibration 160
self-shielding 44, 48
self-shielding halos 41
Sextractor 163, 164
Sheth and Tormen 21

Shull 81
signal-to-noise ratio 162, 186ff
Silk damping 20
simulations 136, 164
size test
    -HII region 109
Sloan Digital Sky Survey *see also*
    SOSS 8, 66, 105
smooth particle hydrodynamics 72
SN-driven winds 96
SNAP 195
solar metallicity 82
Solomon 40, 41
Songaila 84
sources of reionization 124
South Atlantic Anomaly 154
spectral-energy distribution 78, 131
spectroscopy 169
SPH-kernel 72
spherical harmonics 71
spin temperatures 122
spin-flip 121
Spitzer 147, 160
Spitzer IRAC 137
square-kilometer array *see also* SKA
    193
star formation rate 28
star-formation 39, 63, 148
star-formation histories 142
star-formation rate 43, 89, 115
Steidel 87
stellar atmosphere 42
stellar UV radiation 76
Stiavelli 79
Strehl ratio 190
Stroemgren sphere 53, 55ff
super bias 155, 156
super dark 155
supernova 4, 40
    -explosions 9
    -pair-instability 34
SuperNova acceleration probe 195
supernova explosion 49, 56, 57
surface brightness 77ff, 83
surface density 143
Sutherland and Dopita 58
synthetic stellar population models
    131

t
Tegmark 23
temperature of the IGM 175
TFI 185
TFIT 163
thermal equilibrium 11, 121
thermonuclear yields 62
Thompson cross section 12
Thompson optical depth 83, 96ff,
    195
Thompson optical length 13
Thompson optical thickness 18
Thomson cross section 31
Thomson opacity 82
TMT 189
Treecodes 71
Trenti and Hut 70

u
UDF 92, *see also* HUDF 142
UV background 28
UV radiation 41
UV spectrum 80

v
Veron and Veron 153
Virgo consortium 167
virial theorem 22
VLA 153
Vlasov's equation 201

w
WDM 37
White 156
wide-field camera 181
wide-field planetary camera 133,
    185ff
Wise and Abel 63
WMAP 2, 13ff, 95
WMAP optical depth 123
Wright 185

x
X-rays 76